SpeedPro Series

OPTIMISING CAR PERFORMANCE MODIFICATIONS

Simple methods for measuring engine, suspension, brakes and aerodynamic performance gains

More from Veloce:

SpeedPro Series
4-Cylinder Engine Short Block High-Performance Manual – New Updated & Revised Edition (Hammill)
Aerodynamics of Your Road Car, Modifying the (Edgar and Barnard)
Alfa Romeo DOHC High-performance Manual (Kartalamakis)
Alfa Romeo V6 Engine High-performance Manual (Kartalamakis)
BMC 998cc A-series Engine, How to Power Tune (Hammill)
1275cc A-series High-performance Manual (Hammill)
Camshafts – How to Choose & Time Them For Maximum Power (Hammill)
Competition Car Datalogging Manual, The (Templeman)
Custom Air Suspension – How to install air suspension in your road car – on a budget! (Edgar)
Cylinder Heads, How to Build, Modify & Power Tune – Updated & Revised Edition (Burgess & Gollan)
Distributor-type Ignition Systems, How to Build & Power Tune – New 3rd Edition (Hammill)
Fast Road Car, How to Plan and Build – Revised & Updated Colour New Edition (Stapleton)
Ford SOHC 'Pinto' & Sierra Cosworth DOHC Engines, How to Power Tune – Updated & Enlarged Edition (Hammill)
Ford V8, How to Power Tune Small Block Engines (Hammill)
Harley-Davidson Evolution Engines, How to Build & Power Tune (Hammill)
Holley Carburetors, How to Build & Power Tune – Revised & Updated Edition (Hammill)
Honda Civic Type R High-Performance Manual, The (Cowland & Clifford)
Jaguar XK Engines, How to Power Tune – Revised & Updated Colour Edition (Hammill)
Land Rover Discovery, Defender & Range Rover – How to Modify Coil Sprung Models for High Performance & Off-Road Action (Hosier)
MG Midget & Austin-Healey Sprite, How to Power Tune – Enlarged & updated 4th Edition (Stapleton)
MGB 4-cylinder Engine, How to Power Tune (Burgess)
MGB V8 Power, How to Give Your – Third Colour Edition (Williams)
MGB, MGC & MGB V8, How to Improve – New 2nd Edition (Williams)
Mini Engines, How to Power Tune On a Small Budget – Colour Edition (Hammill)
Motorcycle-engined Racing Cars, How to Build (Pashley)
Motorsport, Getting Started in (Collins)
Nissan GT-R High-performance Manual, The (Gorodji)
Nitrous Oxide High-performance Manual, The (Langfield)
Race & Trackday Driving Techniques (Hornsey)
Retro or classic car for high performance, How to modify your (Stapleton)
Rover V8 Engines, How to Power Tune (Hammill)
Secrets of Speed – Today's techniques for 4-stroke engine blueprinting & tuning (Swager)
Sportscar & Kitcar Suspension & Brakes, How to Build & Modify – Revised 3rd Edition (Hammill)
SU Carburettor High-performance Manual (Hammill)
Successful Low-Cost Rally Car, How to Build a (Young)
Suzuki 4x4, How to Modify For Serious Off-road Action (Richardson)
Tiger Avon Sportscar, How to Build Your Own – Updated & Revised 2nd Edition (Dudley)
Triumph TR2, 3 & TR4, How to Improve (Williams)
Triumph TR5, 250 & TR6, How to Improve (Williams)
Triumph TR7 & TR8, How to Improve (Williams)
V8 Engine, How to Build a Short Block For High Performance (Hammill)
Volkswagen Beetle Suspension, Brakes & Chassis, How to Modify For High Performance (Hale)
Volkswagen Bus Suspension, Brakes & Chassis for High Performance, How to Modify – Updated & Enlarged New Edition (Hale)
Weber DCOE, & Dellorto DHLA Carburetors, How to Build & Power Tune – 3rd Edition (Hammill)

Workshop Pro Series
Setting up a home car workshop (Edgar)
Car electrical and electronic systems (Edgar)

RAC Handbooks
Caring for your car – How to maintain & service your car (Fry)
Caring for your car's bodywork and interior (Nixon)
Caring for your bicycle – How to maintain & repair your bicycle (Henshaw)
Caring for your scooter – How to maintain & service your 49cc to 125cc twist & go scooter (Fry)
Efficient Driver's Handbook, The (Moss)
Electric Cars – The Future is Now! (Linde)
First aid for your car – Your expert guide to common problems & how to fix them (Collins)
How your car works (Linde)
How your motorcycle works – Your guide to the components & systems of modern motorcycles (Henshaw)
Motorcycles – A first-time-buyer's guide (Henshaw)
Motorhomes – A first-time-buyer's guide (Fry)
Pass the MoT test! – How to check & prepare your car for the annual MoT test (Paxton)
Selling your car – How to make your car look great and how to sell it fast (Knight)
Simple fixes for your car – How to do small jobs for yourself and save money (Collins)

Enthusiast's Restoration Manual Series
Beginner's Guide to Classic Motorcycle Restoration, The (Burns)
Citroën 2CV Restore (Porter)
Classic Large Frame Vespa Scooters, How to Restore (Paxton)
Classic Car Bodywork, How to Restore (Thaddeus)
Classic British Car Electrical Systems (Astley)
Classic Car Electrics (Thaddeus)
Classic Cars, How to Paint (Thaddeus)
Ducati Bevel Twins 1971 to 1986 (Falloon)
How to Restore & Improve Classic Car Suspension, Steering & Wheels (Parish – translator)
How to Restore Classic Off-road Motorcycles (Burns)
How to restore Honda CX500 & CX650 – YOUR step-by-step colour illustrated guide to complete restoration (Burns)
How to restore Honda Fours – YOUR step-by-step colour illustrated guide to complete restoration (Burns)
Jaguar E-type (Crespin)
Reliant Regal, How to Restore (Payne)
Triumph TR2, 3, 3A, 4 & 4A, How to Restore (Williams)
Triumph TR5/250 & 6, How to Restore (Williams)
Triumph TR7/8, How to Restore (Williams)
Triumph Trident T150/T160 & BSA Rocket III, How to Restore (Rooke)
Ultimate Mini Restoration Manual, The (Ayre & Webber)
Volkswagen Beetle, How to Restore (Tyler)
VW Bay Window Bus (Paxton)
Yamaha FS1-E, How to Restore (Watts)

Expert Guides
Land Rover Series I-III – Your expert guide to common problems & how to fix them (Thurman)
MG Midget & A-H Sprite – Your expert guide to common problems & how to fix them (Horler)

www.veloce.co.uk

First published in August 2018 by Veloce Publishing Limited, Veloce House, Parkway Farm Business Park, Middle Farm Way, Poundbury, Dorchester DT1 3AR, England. Tel +44 (0)1305 260068 / Fax 01305 250479 / e-mail info@veloce.co.uk / web www.veloce.co.uk or www.velocebooks.com.
ISBN: 978-1-787113-18-3 UPC: 6-36847-01318-9.
© 2018 Julian Edgar and Veloce Publishing. All rights reserved. With the exception of quoting brief passages for the purpose of review, no part of this publication may be recorded, reproduced or transmitted by any means, including photocopying, without the written permission of Veloce Publishing Ltd. Throughout this book logos, model names and designations, etc, have been used for the purposes of identification, illustration and decoration. Such names are the property of the trademark holder as this is not an official publication. Readers with ideas for automotive books, or books on other transport or related hobby subjects, are invited to write to the editorial director of Veloce Publishing at the above address. British Library Cataloguing in Publication Data – A catalogue record for this book is available from the British Library. Typesetting, design and page make-up all by Veloce Publishing Ltd on Apple Mac. Printed in India by Replika Press.

OPTIMISING CAR PERFORMANCE MODIFICATIONS

Simple methods for measuring engine, suspension, brakes and aerodynamic performance gains

Julian Edgar

Contents

1. **Testing your car modifications on the road** 5
   Road testing approaches 6
   Road testing places 7

2. **Plotting engine power & torque curves** 9
   Starting points 10
   The instrument 10
   The testing 11
   Example testing 12

3. **Measuring on-road performance** 18
   Stopwatch times 18
   Turbo boost 22
   Gear change points 23
   Torque converters and automatic transmissions 24
   Intake air temperature 25

4. **Flow testing intakes & exhausts** 27
   Measuring intake systems 27
   Making measurements 29
   Total pressure drops 29
   Finding and fixing the problems 30
   Exhaust systems 31
   Fixing problems 34

5. **Testing suspension & brakes** .. 36
   Handling terms 36
   Analysing handling 37
   Skid pan testing 38
   Subjective ride analysis 40
   Measuring suspension natural frequencies 40
   Testing brakes 43

6. **Testing aerodynamics** 45
   Seeing flows 45
   Types of airflow 45
   Wool tufting to test modifications 47
   Measuring aerodynamic pressures 49
   Bonnet (hood) vents 49
   Scoops 50
   Testing lift and downforce 51

7. **Programmable engine management** 55
   On-road test and tune tools 55
   Measuring power when tuning ... 58

8. **Performance modifications: a personal approach** 60
   Intake systems 61
   Exhaust systems 63
   Engine management 64
   Brakes 65
   Suspension 66

Index 72

Chapter 1
Testing your car modifications on the road

I have been testing my car modifications on the road nearly as long as I have been modifying cars. Testing allows me to see what modifications are needed, and, once modifications have been made, whether they work. Over the years I have tested literally hundreds of modifications to engines, transmissions, suspension, brakes and aerodynamics.

Why test on the road? Well, the primary reason is that it doesn't cost anything (or not much anyway). A second reason is that you can do as many tests as you like, and a third reason is that you can do the tests in the way you want them done. A fourth reason is that you can do tests that would otherwise be impossible to do in anything but the most expensive testing facilities in the world.

That might all sound a bit exaggerated, so let's look at each of these points in turn.

Where possible, when I am modifying a car, I test every step of the way. So, for example, I might have decided to make some bolt-on changes to the car that will improve power. Say, a new intake and exhaust.

But, before I lay a spanner on the car, I test the flows of the standard intake and exhaust. These tests are easily done on the road. I've measured the intake systems of some cars and found them extraordinarily bad – as in, it's immediately obvious that a major power gain is available at low cost. But I've also measured other cars where the intake system is so good that it's going to be near-impossible to get any more power by modifying it. Much the same applies to exhausts (although, in my experience, exhausts are typically poorer-flowing than intakes).

If the intake system flows poorly, the next step is to work out where the major flow restrictions are. Sometimes that can be done by inspection and the use of a tape measure, but other times it will need further on-road flow testing. Then, when I have made some modifications to the intake, I test what the result actually is. Say the snorkel (intake to the airbox) looks very restrictive, and the system as a whole flows poorly. I'll then modify or change the snorkel and then hit the road for another five-minute flow test.

Taking this approach means you never do a pointless modification – or at least, it's vastly less likely that you will do a pointless mod! When testing is free and takes little time and effort, why wouldn't you do a lot of it?

The third point was about doing tests as you want them done. Let's say you have a turbo car (I like all types of cars, but over a long time, I've really loved turbo stuff) and you wonder how good the intercooler is. You could take it to a workshop but they will either (a) tell you the

SPEEDPRO SERIES

Measuring information on the road allows you to see what is really going on – not what is assumed to happen. When I first started to measure intake air temperatures on a turbo car, I was amazed at what I found. Controller/displays like this are now cheap and work very well.

intercooler is no good and you should upgrade it – and they just happen to have the right 'cooler to do the job, or (b) they'll put it on a dyno (rolling road) and do a power run, monitoring the intake air temperature, probably via the OBD port or a temporarily added sensor. Unfortunately, neither approach is very helpful.

And what can you do instead? You can fit a temporary gauge or a permanent gauge (my preference), or monitor the OBD port. *And then you can watch that temperature for the next two weeks as you drive in all sorts of conditions!*

When I first did this about three decades ago, I was amazed by the messages the gauge gave me. Having read all the books, I was sure that maximum intake air temp would occur at peak boost – adiabatic compression of the intake air, and all that. But not a bit of it. Maximum intake air temps actually occurred in my cars when stuck in city traffic on hot days, air-conditioner going full blast (and so lots of hot air being driven into the engine bay), and just a trickle of combustion air getting past the almost closed throttle. It wasn't at all uncommon to see 75°C (~170°F) intake air temp in these conditions – and I don't think I ever saw over 50°C (~120°F) with the car moving, even under full boost. The idea of having an intercooler water spray triggered just from intake air temp quickly showed itself to be unviable …

And then it got more puzzling. I had a car with a really big factory intercooler that I used to test going up a long, steep hill at high speed. Peak intake air temp invariably occurred *after the top of the hill was passed!* What was happening was that the intercooler was acting as a heat storage unit as well as a radiator, and when the boost event was over, the intercooler dumped stored heat back into the engine intake. Together with the lower flows, this caused the intake air temp to be higher.

Neither of these things matched what I'd read – what was actually happening on the car was much more complex. And so, when I developed an electronically-controlled intercooler spray, I had the micro-controller monitoring intercooler core temperature as well as intake air temperature (and injector duty cycle for the load input). The spray controller worked very well.

The final point is that you can do tests that would be impossible to perform in anything but the most expensive testing facilities in the world. Here I am thinking of aerodynamics testing. Using simple on-road techniques, you can accurately measure aerodynamic downforce and lift, something that would otherwise require that you have access to a mega-money wind tunnel. You can also find areas of attached and separated flows, test airflow through radiators and oil coolers – and lots more. I'll have more detail on this in Chapter 6.

In short, testing allows you to see what is needed, and then assess the quality of the results your modifications are achieving. And it's easy and cheap.

ROAD TESTING APPROACHES

Unfortunately, it's dead easy to do really *poor* on-road testing that gives you meaningless – or even deceptive – results. To achieve good results, you need to be careful, thorough, organised and dispassionate. Let's look at the fundamentals that will make your testing useful.

1) Design the test for the purpose

Think carefully of what information you are trying to find out through doing the test, then devise a test to achieve that.

Sometimes that's obvious. For example, if you want to find out how well the whole exhaust flows, measure the back-pressure by using a pressure gauge connected to the front of the exhaust. The higher the back-pressure, the worse the flow.

However, sometimes it's not so obvious. If you are trying to measure engine responsiveness in a manual transmission turbo car, you might decide to measure a rolling time (eg 80-120km/h, 50-70mph), using a speed split that involves a gearchange. Or, alternatively, you might decide to drive along at a constant speed, and measure how long it takes to get to a speed that's 10km/h (mph) faster after you mash your foot to the floor.

Note how the 'gear change' method will also be measuring the effectiveness of a blow-off (recirculating) valve, but the other test will not. (Why? Because the

TESTING YOUR CAR MODIFICATIONS ON THE ROAD

action of the blow-off valve will influence spool-up time after a gear change, but won't have any effect on the response after a simple throttle opening.)

2) Don't take short cuts
When testing, it's very tempting to take short cuts that speed up the process.

For example, you always need to do two-way testing. So, when you're making a measurement, you drive in one direction down the road. Then when you've done that, always turn around and do another measurement in the opposite direction. The result is the average of the two tests. This approach takes into account the gradient and wind. However, it can become very tempting to skip one of the tests – you know, both directions are giving the same results, and so ...

Or you're doing rolling acceleration times where you accelerate from one speed, start the stopwatch at another speed, and stop the stopwatch at a third speed. To be consistent and effective, all the tests must be done in just the same way. If you're going faster for that initial speed, abandon the test and start again.

3) Keep good notes
In some testing it's not hard to collect lots of data points. If you just scribble these down as you go along, invariably when you get home you'll wonder what all the gibberish means!

Recently, I was testing the best attack angle of a rear wing to give maximum downforce. I was doing two-way runs at two speeds, so each run was giving me four bits of information. I also trialled eight different wing positions, and so soon had 32 numbers written down. You simply cannot have 32 numbers on a bit of paper in some random order and later make any sense of it. It might sound like overkill, but at least draw up a table that you can fill in as the testing proceeds.

4) Believe the testing data, not the theory
Many decades ago my father told me an interesting story. A defence research scientist, he had been working on a project for detecting atmospheric nuclear bomb tests. He'd developed a theory of how these tests could be detected, and had written an extensive briefing paper on the topic. He was given the go-ahead to perform an experiment proving the theory – and found it didn't work. Not even close. Over the years in working with top engineers, I have heard that again and again. Do careful tests and see what *really* happens.

I said above that I was testing a rear wing to find what angle gave best downforce. I started off with a nice negative angle of attack – it certainly looked right. However, the results weren't anything great. I went to a higher angle of attack – still nothing good. I went still steeper – no good. Then, becoming quite puzzled, I tried a positive angle of attack – that is, the front of the wing *higher* than the back. The wing started to work! Why? Because the airflow over the sloping rear hatch of the car was partly angled downwards – and so, compared with true horizontal, the wing needed a positive angle to be effective. Testing showed what *really* worked …

ROAD TESTING PLACES
The roads on which you test are probably going to be the roads near where you live. The emptier the roads, the better. With empty roads you can vary speeds, go really fast and really slow, you can stop more frequently, and you can more easily

---

Surface pressure measurements, centreline

Car: Mercedes E500 W211
Date: 13/11/2017
Test road: North of Dalton
Weather: Fine, 10 km/h east wind, 23°C
Equipment: Magnehelic, surface disc probe, 70 km/h

Location	Direction 1	Direction 2	Average
Front number plate	+120	+130	+125
Leading edge of bonnet	-130	-100	-115
Base of windscreen	+90	+70	+80
Top of windscreen	-140	-90	-115
Top of rear window	-70	-30	-50
Base of rear window	-10	~0	-5
Centre of boot lid	-20	~0	-10
Rear numberplate	-40	-10	-25

Reminder: reset reference volume pressure to atmospheric before each run

These are the sort of notes I keep when testing. This particular sheet is for aerodynamic testing of surface pressures (not covered in this book – see my book *Modifying the Aerodynamics of Your Road Car*, also published by Veloce) but I take a similar approach for all on-road testing.

SPEEDPRO SERIES

perform aerodynamic tests like those involving wool tufting. Where I live now, in rural Australia, I can use roads that I can be fairly certain will be empty of other traffic for 10 or 15 minutes at a time. But I also have lived in major cities teeming with traffic. In those cases, I did testing in industrial areas at night; I did testing on freeway 'on' ramps (especially good for rolling acceleration times!) and I did testing on roads that I had to drive for quite a while to get to. If you have roads on which you can drive, you can test.

However, when you are testing, other people on the road won't know what you are doing, and some people – including the police – will get quite unhappy with you if you go about it the wrong way. For example, testing of top speed is a technically valid approach when measuring aerodynamic downforce or lift, aerodynamic drag, aerodynamic stability, full-load air/fuel ratios, full-load engine oil temperatures and the like. Trouble is, unless you live in just a few places in the world, you'll potentially be in a lot of trouble if you are caught doing this.

(An aside: I once appeared in court on a charge of doing 150km/h in a 110km/h zone. I was testing the changes in the aerodynamic stability of my turbo Toyota Prius caused by fitting a front undertray. In court all I could think of saying was the truth. I still remember the rustle of surprise in the court room when I said what I'd been doing. I was lucky – I received just a fine.)

Despite what I have just said about high-speed testing (and it's not something I recommend you do – if you kill someone you could be jailed), I think that's the only brush with the law I have had in doing literally thousands of on-road tests.

If you have a very powerful car, there will be a greater limit to what you can do on-road. However, the vast majority of techniques covered in this book will still be valid. If you have a lower powered car, then it becomes easier, as, for example, full-throttle testing will be able to be carried out over a wider range of engine revs.

But whatever your car and road availability, you can do a lot of free and effective testing. So let's get straight into it!

Not everyone has stunning and empty roads like this to test on, but most testing can be done on local roads, even if you live in a city. (Courtesy Georgina Edgar)

Chapter 2
Plotting engine power & torque curves

One of the big changes in car modification over the last few decades has been the increasing popularity of chassis dynamometers. These days, many workshops have dynos (rolling roads) where you can drive onto the rollers and quickly measure engine power output over the rev range. This not only shows peak power but also the shape of the power curve. Small gains or losses can be easily seen, allowing you to quickly assess the results of tuning changes or performance modifications.

However, for the home modifier, there are a few problems with dynos. The greatest is the cost. Unless you're very good friends with the workshop, the cost of a dyno run rules out doing one after each small modification has been made. And, if you're tuning programmable management yourself (another good use of a dyno), you probably won't want to pay the hourly rate for using a dyno while you tune your car.

However, there are some other downsides of dynos that people seldom mention. One is that the environment in which the engine is developing power is not realistic. The airflows around the car, the rate at which the engine accelerates, and the intake air temperatures are often quite different to that found on the road. On a turbo car, for example, it's impossible to replicate real-world intercooler efficiency.

So, is there a way of measuring, on the road, the shape of the engine's power and torque curves, especially so that gains and losses in power can be seen? There is, and it's simple and cheap. It doesn't replace every function a dyno can provide – but on the other hand, it has the benefit of being done in realistic conditions. Note that this approach will not tell you the amount of horsepower or kilowatts

A dynamometer is a very useful tool, but it has major downsides for someone modifying their own car. In addition to the high cost of every run, the dyno cell also does not replicate on-road conditions of airflow and temperature.

SPEEDPRO SERIES

your engine is producing, but it will allow you to accurately see *changes* in power and torque, and see the *shape* of these curves. In turn, that allows you to see the results of changing cams, changing turbo size – even the effect of different turbo boost control systems.

STARTING POINTS

Let's first take a step backwards to understand what we're trying to measure. We will start off with engine torque, and get to power later.

When a car travels down a road, its tyres are pushing back on the pavement. If the push backwards equals the forces the car needs to overcome (on a flat road that's aerodynamic drag and rolling resistance), then the car will move at a constant speed. If the force pushing backwards (called the tractive effort) is greater than the forces that need to be overcome at that moment, the car will accelerate. The greater the surplus of tractive effort, the greater the acceleration.

If the greater the acceleration, the greater the tractive effort; and the greater the tractive effort, the greater the torque being produced by the engine, then by measuring actual on-road acceleration we can see the shape of the engine's effective torque curve. Taking this approach then automatically takes into account losses that occur at different rpm and gear-train loadings, losses due to accelerating rotational masses – the whole lot. It's a measurement of the available torque at the wheels to cause acceleration at that speed in that gear.

There's no need for a dyno or expensive gear – and in fact, it's more accurate than a dyno because it is like a dyno with an infinitely big single roller (well, one as big as the diameter of Earth!), and like a dyno mounted in a variable speed, climate-controlled wind tunnel. Furthermore, this incredible dyno is also programmed for the actual, varying 'ramp rates' that occur when your car accelerates in each gear!

I said above that the greater the acceleration, the greater the tractive effort (and so engine torque) being developed at that point in time. So, how do you measure acceleration at a given point in time? This is called instantaneous acceleration, and in fact, it's very easy to measure.

THE INSTRUMENT

Instantaneous acceleration is measured with an instrument called an accelerometer. Performance

An app for a smartphone that will read out acceleration in one plane can be used to measure instantaneous, on-road acceleration.

measuring accelerometers are available in two types – electronic and mechanical.

Electronic accelerometers are most easily obtained by using a 'g-force' application in a smart phone. All you need is an app that shows actual g-force in big numbers on the screen as it occurs real time. Be careful though, because you want to measure only longitudinal acceleration. Make sure that the app does in fact do this, and doesn't also try to simultaneously measure vertical and lateral accelerations.

An alternative is a mechanical accelerometer. A mechanical accelerometer can use either a vertical pendulum that is deflected across a scale by acceleration, or a tube shaped in a semi-circle in which a small ball bearing is moved. A US company called Analytical Performance some years ago produced one of the best of the latter type of accelerometers, which was called the G-Curve. Their accelerometer consisted of an engraved alloy plate into which was let a long, curved glass tube. The tube was filled with a damping fluid and a small ball bearing was sealed

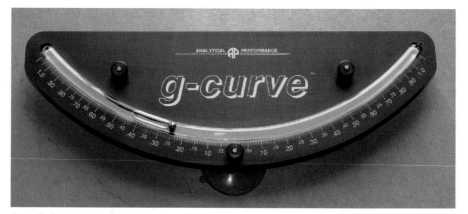

The G-Curve from Analytical Performance introduced me to accelerometers. This is a superb instrument that can accurately measure instantaneous acceleration. Note the ball that you can see in the tube – this climbs the incline as the car accelerates. From this data you can plot engine torque and power curves. The G-Curve is no longer available, but you can use other instruments to achieve the same outcome.

PLOTTING ENGINE POWER & TORQUE CURVES

A boat angle of heel indicator can be used in the same way as the G-Curve. As shown here, you will need to make a mount to allow it to be positioned vertically on a side window. The device is calibrated in degrees but these are easily converted to acceleration, measured in g.

inside. A very good handbook was also provided with the instrument.

Unfortunately, Analytical Performance is long out of business, but a substitute accelerometer can easily be assembled. Boat and yachting supply companies sell clinometers that are designed to measure the angle of boat heel. One such clinometer is the 'Lev-O-Gage', which in construction is very similar to the G-Curve. However, because it is designed to measure heel angles, the scale is calibrated in degrees rather than g units. Like the G-Curve, the glass tube is filled with a damping fluid to prevent the ball overshooting.

Both mechanical and electronic accelerometers are attached to the car in the same way. To measure longitudinal acceleration, the instrument is mounted level and parallel with the direction the car is moving. This means that the accelerometer is often mounted on the passenger side window. The G-Curve came with suction caps to allow this to be easily done, but the Lev-O-Gage is designed to be fixed in place with double-sided tape, and so a suitable bracket should be made and then equipped with suction caps (available from rubber supply shops). Lev-O-Gage also sells a suction cap assembly with an adjustable bracket that looks like it would work well.

Let's concentrate now on the Lev-O-Gage. When the car accelerates, the ball climbs up one arm of the curved tube, showing how hard the car is accelerating. To convert the degrees reading of the clinometer to g readings, simply use a scientific calculator to find the tangent (tan) of the number of indicated degrees. This means that if the car is accelerating hard enough to move the ball to the 20-degree marking, the acceleration is about 0.36g (tan 20 = 0.3639). However, note that converting the degree readings into 'g's isn't necessary for most testing, where you are only trying to see changes rather than measure absolute values.

Both the electronic and mechanical accelerometers must be accurately levelled before performance measurement can take place. The accelerometer will clearly show any road gradient, and so levelling is best done on a flat road. If the road is not level (and many ostensibly level roads actually aren't), the instrument needs to be adjusted so that the error when the car is parked facing in opposite directions on the same spot is of the same amount but in different directions. For example, the clinometer might be set to show +2° with the car facing in one direction and -2° with it facing in the other. If the road was level, the instrument would therefore show 0°.

Note that using either an electronic or mechanical accelerometer also requires a sharp-eyed assistant armed with a paper and pencil to record the data.

THE TESTING

So how do you do the testing? A gear is selected, and the car driven at as low a speed as is possible in that gear. Warn your assistant that you are about to start the run by saying "Go!" and then quickly push the accelerator to the floor. At every 1000rpm of engine speed, yell "Now!" Each time you yell "Now!" your assistant records the accelerometer reading. On cars

Accelerometer readings can be plotted against either engine speed (rpm) or road speed. In this chapter I mostly use rpm, while, in the next chapter, readings are plotted against either rpm or road speed.

SPEEDPRO SERIES

with low engine redlines, you can make the readings every 500rpm. Depending on the gear in which you test and the performance of the car, the 'g' readings might be quite small (eg 0.15g) or much larger (0.6g).

In many cars, the acceleration will be too quick for the assistant to keep up, so on the first run do for example 2000, 4000 and 6000 rpm, and in the second run do 3000, 5000 and 7000 rpm. You can also have the assistant yell out the readings (quicker than writing them down) and record it on a smartphone. The taller the gear, the slower will be the acceleration. In cars with automatic transmissions, you will need to manually select and hold the gear, ensuring that kickdown does not occur at full throttle. For greatest accuracy, make a run in each direction and then average the readings.

You'll then end up with a table of readings of instantaneous acceleration (g) and engine speed (rpm). Use Excel to graph the data, selecting a smoothed line. (If you don't know how to do this, just plot the graph by hand on graph paper.)

EXAMPLE TESTING

Let's look at some example measurements made with an accelerometer. The following table shows the measurements made in 3rd gear acceleration on a tiny 660cc turbo Daihatsu car:

Engine speed (rpm)	Acceleration (g)
2000	0.14
3000	0.2
4000	0.21
5000	0.2
6000	0.15
7000	0.12
8000	0.09

In table form the figures don't seem to mean much, but that all changes when we graph the data. Figure 2-1 shows this. Now it makes much more sense, because this is the engine's effective torque curve.

Let's look at this in more detail. The amount of twisting force the crankshaft can develop is dependent on the pressure in the cylinder – and off-boost, this little turbo engine doesn't have much! However, between 2000 and 3000 rpm, the engine comes on boost and so torque rises rapidly. The boost control then comes into effect, holding boost at a fixed level from 3000 to 5000rpm, but then the torque starts to fall away as engine breathing becomes the restricting factor at high revs.

Now let's modify the car. A big exhaust was fitted and boost was increased.

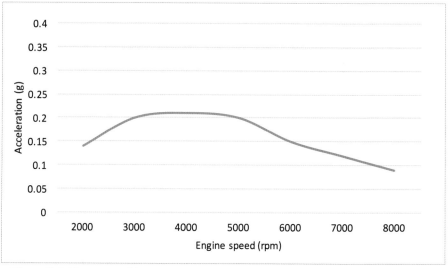

Figure 2-1: The on-road instantaneous acceleration, measured at each 1000rpm of engine speed, shows the shape of the engine's torque curve. This turbo engine holds a flat torque curve across the mid-range, but then falls away as breathing is restricted at higher revs.

REALLY POWERFUL CARS?

If you have a really powerful car, you may find that either things happen too quickly for the technique described above to work, or you need to be in too high a gear, and road speeds are therefore excessive. In that case, you will probably need to log the information electronically. You need two sets of data – measured instantaneous acceleration and engine rpm.

If you have programmable engine management or an aftermarket electronic dashboard fitted to your car, it's likely that these two sets of data can be logged and then played back. For example, in one of my cars, the MoTeC dash, can log both longitudinal acceleration and engine revs.

If you don't have a programmable electronic dash or similar, video the accelerometer and have the driver call out each 1000rpm. Played back slowly – or with frequent pausing – will allow you to access the acceleration and rpm data.

PLOTTING ENGINE POWER & TORQUE CURVES

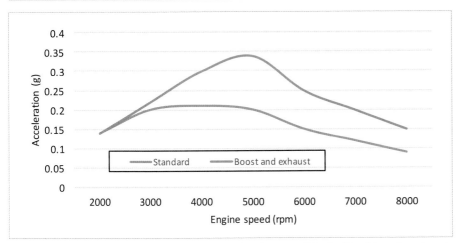

Figure 2-2: The same turbo engine as Figure 2-1, but now fitted with a big exhaust and with boost increased. You can see that, once the engine comes on boost, instantaneous acceleration (and so engine torque) is up everywhere, but that the engine is more peaky.

Engine speed (rpm)	Standard acceleration (g)	Modified acceleration (g)
2000	0.14	0.14
3000	0.20	0.22
4000	0.21	0.30
5000	0.20	0.34
6000	0.15	0.25
7000	0.12	0.20
8000	0.09	0.15

Again, at a glance the figures don't mean a lot, so look at Figure 2-2. You can see that at low revs, there is no change in torque – as you'd expect with increased boost and a large exhaust. But once the turbo comes on stream, the higher boost level immediately makes its presence felt, with torque continuing to grow until it peaks at 5000rpm. In fact, at 5000rpm, the actual acceleration in 3rd gear has risen by 70 per cent! The higher torque level hangs on – although not quite to this extent – right to the redline.

However, and here's where the worth of this measuring technique really shows itself, you can also see something else. The engine is now much peakier. Put your foot down at 3000 rpm in 3rd gear and you get 0.2g acceleration, but do the same at 5000rpm, and you get 0.34g acceleration! Compare that with the standard car where the acceleration at 3000, 4000 and 5000rpm is much the same. To put this another way, when cornering hard on the throttle, the modified car will be much harder to drive because the torque delivery has become quite non-linear.

Now let's put a bigger turbo on the car. The following table shows the on-road measurements.

Engine speed (rpm)	Standard turbo, boost and exhaust (g)	Big turbo, boost and exhaust (g)
2000	0.14	0.12
3000	0.22	0.13
4000	0.30	0.15
5000	0.34	0.23
6000	0.25	0.34
7000	0.20	0.25
8000	0.15	0.23

WHAT'S A G?

So far I've been talking a lot about measuring acceleration in g units. But what's a g?

'G' stands for gravity, and a unit of 1g indicates the acceleration due to Earth's gravity. We're all attracted towards the centre of the Earth, and if we were free to fly in that direction, we would accelerate towards the centre at a specific rate. That rate is 9.8 metres/second *per second.* In other words, our speed would get faster by 9.8 metres/second *each second* we were accelerating. So, each second, our speed would *increase* by about 35km/h, or about 22mph. You can see that 1g is quite a strong acceleration – as we find out if we trip and fall over.

One advantage of measuring in g units is that it's easy to make mental comparisons with weights. That's because our weights weigh what they do because of Earth's gravity. So what do I mean?

In Chapter 5 I'll show you how to calculate lateral acceleration (again measured in g units) by timing your speed around a skidpan. Let's say that you are pulling 0.5g lateral acceleration and your car weighs a neat 1000kg (1 tonne). The sideways force developed in total by the four tyres is therefore half of 1000kg, or 500kg. To put that another way, let's attach a spring balance to the car's centre of gravity and pull sideways with a force on the balance of 500kg – that's the exact force in action in a 1 tonne car cornering at 0.5g.

Or let's say that you're braking at 1g. In that case, in a 1-tonne car, the rearwards force developed by all four tyres adds up to 1000kg. A 1g vertical bump? The load on the springs doubles.

Using g units allows lots of things to be much more easily pictured than working in m/s/s or ft/s/s.

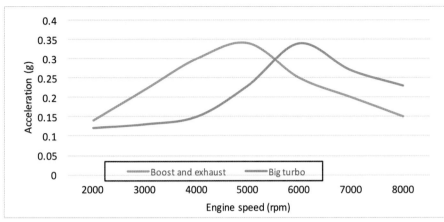

Figure 2-3: The measured instantaneous acceleration curves, with standard turbo (big exhaust and more boost) versus a new, larger turbo. You can see that peak torque has been moved a long way up the rev range, and that for most of the rev range, performance is worse than standard.

Figure 2-3 graphs the data. Oh my gosh, what has happened? You can see that with the bigger turbo there is very little torque below 4000rpm, and in fact to develop the same on-road acceleration as was achieved with the little turbo takes until no less than 5500rpm! It looks like a bit of a disaster – and to be honest, on the road it was a rather like that. But if you look at the power curves, you'll see a quite different story.

So let's now turn to power. Power is simply torque multiplied by engine speed, and since we're using instantaneous acceleration as an indicator of torque, all that we need to do is to multiply the measured acceleration by the rpm. Let's do that for the last table we were working with (see below). To keep the numbers manageable, we'll divide engine speed by 1000 first – ie 2000rpm becomes 2.

Figure 2-4 shows these results. (Note that we're not assigning units to power – they're relative numbers.) When looking at the power curve you can see that with the big turbo, power has increased quite nicely – in fact, it's up by over 50 per cent at 8000rpm. So if you were the sort of person who liked driving everywhere at 6000-8000rpm, this would be a pretty good outcome. However, in the real world, it doesn't look like a good upgrade – and it wasn't.

(The curves shown in Figure 2-4 are symptomatic of many modified turbo cars that have had big turbos fitted. The owner says excitedly "Now it's got 500hp!" (or whatever) but in actual fact, the average power available across the rev range is probably less than standard. Measured on-road torque curves show this well.)

In the data we've looked at so far, the changes in performance have been quite major. But the beauty of this approach is that it can also measure small changes – and so find modifications that don't in fact actually work.

A few years ago, I owned a Skoda Roomster 1.9 diesel turbo. It was a very roomy and frugal car – albeit not very fast – and I became intrigued in giving it a bit more power without wrecking the fuel economy. Using the techniques covered later in Chapter 4, I measured the intake system restriction and found that it was a poorly-flowing design. I also looked around and found a lot of discussion among owners of cars with this engine (it was widely fitted to Volkswagen group products) suggesting that a big intake made a noticeable difference to performance.

I therefore decided to build a free-flowing intake for the car, but before I started, I measured the on-road instantaneous acceleration across the rev range. I then built the intake – and I couldn't feel any improvement on the road. I then re-measured the on-road instantaneous acceleration. The 'before' and 'after' measurements showed that except at low and very high revs, the standard car had more torque than the modified car! So if anything, the intake had made things *worse* – Figure 2-5 shows this.

In fact, a similar story was

Engine speed (rpm)	Standard turbo, boost and exhaust (g)	Power	Big turbo (g)	Power
2000	0.14	0.28	0.12	0.24
3000	0.22	0.66	0.13	0.39
4000	0.30	1.20	0.15	0.60
5000	0.34	1.70	0.23	1.15
6000	0.25	1.50	0.34	2.04
7000	0.20	1.40	0.25	1.75
8000	0.15	1.20	0.23	1.84

PLOTTING ENGINE POWER & TORQUE CURVES

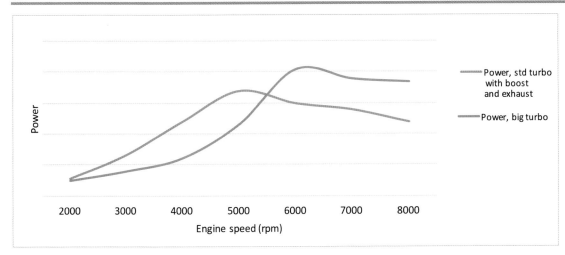

Figure 2-4: The power curves for the standard and big turbos. Note how peak power has increased by a lot – in fact, it's up by 50 per cent at 8000rpm. However, power is also down over a lot of the rev range – very important unless you want to drive around at 6000-8000rpm all the time.

repeated when I added a new, free-flowing muffler and a much larger intercooler. Performance varied little over standard – and there was certainly no clear-cut gain. The problem was that a diesel is heavily reliant on the mass-flow of fuel into the engine – and without more fuel being added, there was no more power. However, when I had the car's ECU re-mapped, the car responded very well indeed. Figure 2-6 shows the resulting change in the torque curve and Figure 2-7 shows the calculated power curves. As can be seen, the end result of all the modifications was very successful. The final gain was so spectacular (standard turbo, remember) that I am sure the results would have been worse without the interim modifications.

And a final example. By using a two-stage control system on an Audi S4, I was able to lift mid-range acceleration by 17 per cent, a major improvement in mid-range torque. And this was with identical before/after peak boost – the boost wasn't 'turned up' at all! Figure 2-8 shows the outcome.

From these examples I hope that you can see a couple of things. Measuring on-road

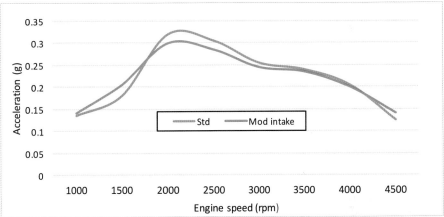

Figure 2-5: The result of fitting a free-flowing intake to a 1.9-litre turbo diesel. Over most of the rev range, performance was worse! Without increased fuelling on this diesel engine, the mod was pointless.

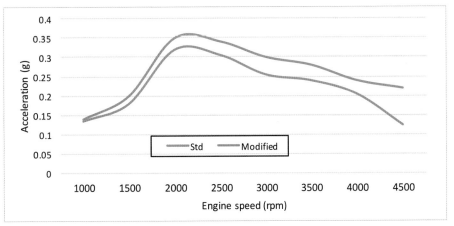

Figure 2-6: The measured torque curves, with the combination of new intake, rear muffler, intercooler and custom ECU re-map. Acceleration is better everywhere in the rev range.

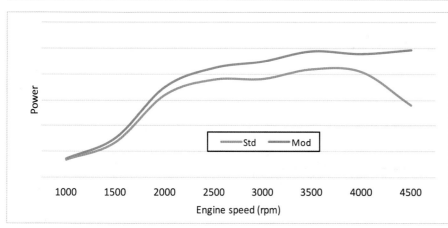

Figure 2-7: The calculated standard and modified power curves resulting from the new intake, rear muffler, intercooler and custom ECU re-map. The extraordinary gain at high revs was very much able to be felt on the road.

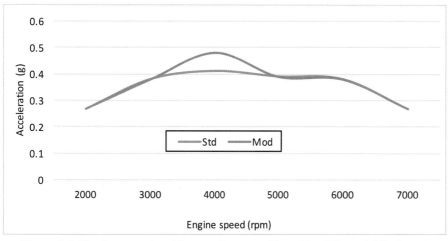

Figure 2-8: The increase in mid-range torque achieved by fitting a tricky boost control to an Audi S4. Peak boost was unchanged.

instantaneous acceleration gives you all the information you need to decide if your modifications are heading in the right direction. On a road car, where you want to use a wide range of engine revs, making modifications that improve instantaneous acceleration across the whole rev range gives the best results. If this is not possible, ensure that you don't go backwards in torque anywhere. Increasing measured acceleration at high revs will make a large difference to developed power, but in the real world, gains in top-end power, without commensurate gains elsewhere, are usually not very helpful.

Over the years, I have used the measurement of on-road acceleration to plot power and torque curves for many of my cars. I think it is a stunningly useful technique. It can be used to measure very small changes (and if you cannot measure the on-road change, there hasn't been enough of a change!) and it does it all in absolutely real-world conditions. Every power-producing modification – from exhausts to engine swaps – can be tested in this way.

In the next chapter, I'll look at making more on-road measurements, including using the accelerometer to determine best gear change points, among other uses.

Any power-producing modification can be measured using the techniques covered in this chapter, including the fitting of a supercharger, as shown here.

PLOTTING ENGINE POWER & TORQUE CURVES

ACCURACY

So how accurate is the approach that has been covered here?

First, the slower the road speed at which you can make the measurements, the better. This is because as road speed increases, so does aerodynamic drag. Drag is a resisting force, and so if this resisting force is really high, the car will have poor acceleration, even if the available torque is high. Therefore, if you are doing runs in one gear from say 50km/h to 150km/h (about 30-95mph), the curves will be a bit pessimistic at high rpm. That makes no difference at all to comparing performance modifications, though – just always do your testing in the same gear.

Second, and contradicting the above suggestion, is that the higher the gear you can do the testing in, the better! Why? Because in lower gears the rate of acceleration is faster, and so the effects of the rotational inertia of the driveline will be greater. In effect, you are not only accelerating the car, you're also accelerating all the rotating bits. Again, if you are simply making comparisons before and after modifications, this doesn't have an impact – unless of course you are reducing the mass of rotating parts of the car.

Third, if you have the combination of a powerful car and soft suspension, the car will squat under acceleration, and so the measured g-values will be influenced. Yet again, this doesn't matter if you are making comparisons of relatively small changes in power.

So, overall, how does it all stack up? Figure 2-9 shows the on-road measured power curve of the standard 1.9 litre turbo diesel Skoda Roomster, compared with the factory power curve for the engine. I have moved the curves vertically until they appropriately overlap – remember, we're comparing only the shape of the curves. As can be seen, the shapes are very close – and I'd suggest that the on-road measured data at low revs is actually more accurate than the Skoda factory figures. Why? The Skoda figures would have been measured with an engine dyno. A dyno can better load the engine at low revs than the acceleration of the car; this typically results in more boost at very low revs on the dyno than on the road. Also, the dyno typically doesn't have the ramp rate of the accelerating car, so it doesn't see the slight boost overshoot that the car has at 2000rpm.

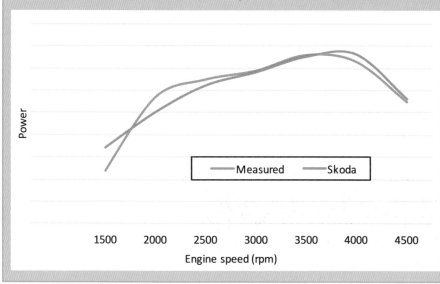

Figure 2-9: Comparing the shape of on-road measured power curve of the Skoda Roomster 1.9 turbo diesel with the Skoda factory power curve. It's likely the on-road measurement is more representative of reality than the factory curve.

Chapter 3
Measuring on-road performance

Talk about measuring car performance on-road, and people's minds invariably turn to tyre-smoking launches as you weave down the road, trying to control the errant beast. However, while standing-start times are part of measuring on-road performance, much performance measurement is a lot less dramatic – and also far more useful.

STOPWATCH TIMES

By far the best tool for assessing vehicle performance is a stopwatch and the car's standard speedo. With care, stopping and starting a hand-held stopwatch can be carried out with a consistency of less than $1/10$ of a second. And it should be noted that any performance mod that results in less than a tenth of a second gain over a reasonable speed increment isn't very successful!

Practise starting and stopping a stopwatch while you view the second hand on an analog clock or the digital seconds display on another watch. Time a number of 10-second increments on your

The humble stopwatch is a very powerful tool when assessing performance modifications. Most people immediately think of standing-start times, but rolling acceleration times are more useful. (Courtesy Jadco)

stopwatch and see how close you actually get to 10.0 seconds. I have just stepped away from the keyboard to do this and have got the following results: 9.94, 10.10, 9.98, 9.92, 9.94 and 9.96 seconds. You can see that when the single figure furthest from the median is excluded, the timing is only 0.6 per cent inconsistent. Obviously, with the g-forces pushing you back in your seat and the speedo needle whipping around while the road races towards you, the in-car timing won't be this accurate, but using a digital stopwatch is far better than most people think.

Using a stopwatch to measure acceleration performance in this way does not require that the speedo be accurate. If you wish to compare your results with figures gained in other cars, then of course the speedo must be correct, but in the vast majority of cases you will be simply comparing the performance before and after

MEASURING ON-ROAD PERFORMANCE

making a modification. Unless you change the tyre diameter, final drive ratio, or some other gearing aspect, the speedo accuracy won't change during this period.

Rolling times

The most important real-world performance times are those measured 'in-gear' or 'rolling.' These can be done in two different ways.

An *accelerating* rolling time is one where, for example, you drive along in second gear at 50km/h. You put your foot flat to the floor and then click the stopwatch 'start' button as the speedo needle moves past 60km/h. You then again click the stopwatch at 90km/h – and then back-off after that. So you're measuring how quickly the car will accelerate from 60 to 90km/h *when it has already begun accelerating.* (A similar increment would be 35-55mph, with the acceleration starting from 30mph.)

The other way of doing a rolling time is to drive along at a constant 60km/h and, at the same time as you move to full throttle, press the stopwatch. Again, stop the watch at 90km/h (or whatever speed you nominate). This measurement not only takes into account acceleration, but also response – the time it takes for the car to begin to accelerate. This type of split is called a *static* rolling time.

Note that the speed increments described here can be used almost anywhere on suitably speed-rated roads – from a freeway 'on' ramp to a country road. The actual speed range you choose to use, and the gears you do it in, are up to you.

These two 'rolling' measurement techniques are exceptionally useful – far more so than the standing-start times I'll cover in a moment. In the real world of road driving, I'd always trade a drop in power for better rolling throttle response. Why? Two reasons. First, by far the majority of driving does not involve accelerating away from a standstill. Second, a faster-responding car will always be much more enjoyable than one with (say) 10 per cent more power but a laggier response. A responsive car is also much easier to corner on the throttle.

As I said, a static rolling time is very good for assessing responsiveness. Turbo folk will immediately think of turbo lag, but responsiveness is also heavily influenced by transient ignition timing and fuelling. So, if you are mapping engine management, measuring this sort of split can be very important. Don't do as I once did and just assume the responsiveness (or lack thereof) is 'just how things are' – I more than halved the time the car took to respond by carefully mapping transients in fuel and ignition.

Talking of a different car, I once did some modification of a diesel Peugeot 405. The car was a pretty slow (but very economical) machine, and so I wasn't expecting to revolutionise performance. But, by making changes to fuelling, the intake and exhaust, I dropped the 'accelerating' 80-100km/h time from 10.7 seconds to 5.5 seconds – and, after the modifications, the car took about half the time to accelerate between these speeds in 4th gear. 80-110km/h in 3rd gear improved even more dramatically – falling from 15.1 seconds to 5.9 seconds. Importantly, the standing start time 0-100km/h difference was much less impressive – but on the road, the car felt transformed. It is also important to note that on the dyno, the measured gains were minor ...

Another test was done on a different car, comparing the supposed power gains of an engine oil additive. Testing was carried out using an 'accelerating' 40-80km/h in second gear. In standard form, the 40-80km/h times were 3.80, 3.82 and 3.86 seconds. I then added the oil treatment which – among other things – was claimed to give up to 15 per cent more power. Figure 3-1 shows the 'before' and 'after' times. Each of the 'after' runs was quicker than the 'before' runs – but the difference was so tiny as to make the additive worthless – at least from a power point of view. Incidentally, the test results of the oil additive didn't surprise me. I've seen similar (non) results from

A Peugeot 405 diesel. By making strategic modifications, the rolling acceleration times over some speed increments were more than halved. This revolutionised on-road performance, but these gains were not reflected in either standing-start times or dyno power measurement.

SPEEDPRO SERIES

Rolling acceleration times are particularly interesting in turbo cars. Comparing the one time split will show that an 'accelerating' rolling time will give much quicker times than a 'static' rolling time. The difference between the times is largely due to turbo lag. (Courtesy Ford)

special sparkplugs, special ignition leads, replacement 'sports' airfilter elements, etc.

I said above that the splits and gears that you use in rolling acceleration times are up to you. But I would add a caveat. Make the conditions ones that you will be using in real life. For example, pick a gear and speed split of the sort that you use when overtaking – eg 80-110km/h (about 50-70mph) in third gear. In urban conditions, 60-90km/h (about 35-55mph) in second gear is a commonly-used combination. Of course, there's nothing to stop you doing a 40-140km/h (25-85mph) split in 4th gear – but when would you ever be in 4th gear at 40km/h (25mph) … not if you wanted to go hard, anyway?

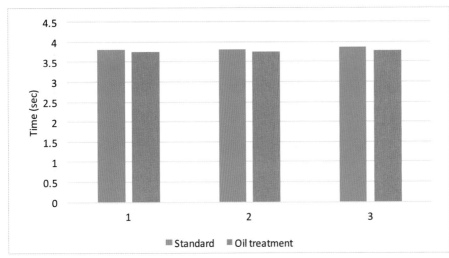

Figure 3-1: The measured rolling 40-80km/h times with and without an oil additive, with three comparison runs made. The additive was claimed to provide 15 per cent more power. It didn't.

DECEPTIVE GRAPHING

Now seems as good a time as any to talk about deceptive graphing. Have a look again at Figure 3-1 (above). You can see from the graph that (a) the variation in times in the same configuration is small (ie the testing is consistent), and that (b) the variation in times, with and without the oil treatment, is also small. And that's largely because the graph, as with every single one in this book (except the following example), starts its vertical axis at zero.

So why is that important? Take a look at the graph here. It shows *exactly the same data*.

But wow! Doesn't it look like the oil treatment does a lot now! Just look – those times after the oil has been treated are so much shorter! But remember, *it's the same data!* All I have done is some manipulation of the data order and the graphing style. Significantly, the graph's vertical axis no longer starts at zero, much exaggerating the difference between the two sets of times. Grouping them into 'before' and 'after' just adds to the effect …

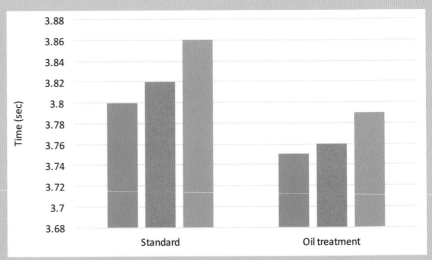

MEASURING ON-ROAD PERFORMANCE

Standing-start

The greatest variation in standing-start acceleration times is that caused by differing launches.

My near-new, standard R32 Skyline GT-R used to get to 100km/h (about 60mph) in low sixes, if launched and driven as you would if you owned the car. However, on a cold night, clutch-dumped at the redline by a maniac (me in a don't-give-a-damn mood) it did this time in the high fours.

A high-powered manual RWD car – and any manual FWD or constant AWD – car is very difficult to launch consistently. In both types of two-wheel drive cars, wheelspin will occur (or the traction control be activated) if too many revs are used on launch, while a constant four wheel drive car will very easily bog down. For this reason, accurately measuring performance changes with these cars is better done from a rolling start. On the other hand, all cars with automatic transmissions lend themselves very well to consistent standing-start timing.

So how do you do the timing? Set the stopwatch to zero (it helps if it beeps when started and stopped) and then rev the engine to the launch rpm. In an automatic car, load the engine with the brakes while applying some accelerator. Release the clutch (or brakes in an automatic car) as you press the stopwatch and then accelerate hard through the gears. Prepare to press the 'stop' button as the speedo needle sweeps around, hitting the button a fraction before you see the needle actually pass over the mark.

As with all types of acceleration timing, if you are after the most accurate figures possible, make three or four runs in each direction. You will quickly see if the figures form a pattern – or are simply all over the place. Don't confuse timing

ELECTRONIC TIMING?

You may be reading this and thinking that many of the techniques look rather old-fashioned. Where's the electronic timing app running in the smartphone? Where's the datalogging from the OBD port?

In short, while I have used all these techniques (and many more), I prefer the simple, effective and very reliable approaches shown here. (Later in this book, when measuring suspension natural frequency, I use a smartphone and an app – there's no better way of doing those particular measurements.)

Using a performance measuring app in a smartphone requires that (1) the software is very well programmed, and (2) the phone remains dead-level during the run. Importantly, there's often also no way of running static and accelerating split times. Logging from the OBD port is fraught with difficulty as the update rate is usually insufficient for the required accuracy. Electronically logging of both rpm (or road speed) and instantaneous acceleration can be done – but you'll typically need a motorsport level ECU or dash. (Or just use a video camera as described in the previous chapter.)

Honestly, if you can't pick performance improvements with a stopwatch or accelerometer, it's because the improvements are so small as to be worthless.

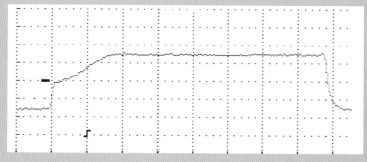

Figure 3-2: This graph shows turbo boost measured in second gear of a car with an automatic transmission, a log made to show how well a newly-developed electronic boost controller worked. I recorded this using an electronic pressure sensor and expensive Fluke Scopemeter – but the same thing could have been done with an analog boost gauge, an assistant, and a pen and paper!

inaccuracies with the performance of the car changing during the testing; as the engine heats up, power (especially in a forced induction car) will often decline.

I carried out some standing start testing on a standard six-cylinder car. The automatic car recorded a 0-100km/h (about 60mph) time of 9.2 seconds in standard trim. The bonnet (hood) was then 'popped' to the safety catch. This let more air get to the factory intake snorkel and the time dropped to 9.1 seconds. The bonnet was then returned to its closed position, and the intake snorkel to the airbox removed, allowing the engine to breathe hot air. The time then lengthened considerably to 9.5 seconds. The engine was fitted with a dual length intake manifold that changed from long to short runners at a certain engine speed. For the final test it was permanently held in its short runner position, resulting in a slow

SPEEDPRO SERIES

9.9 second time. Many years later, I did some further testing on the same type of car in modified, 5-speed manual form. I was trialling variations in cam timing. In that case, the 0-100km/h time did not change with the cam timing in either of two positions – it stayed at 7.5 seconds, and so altering the cam timing made no difference to power.

These days, I am very reluctant to perform standing-start measurements in any manual transmission cars. The amount of information you gain, compared with the wear and tear on the car, is not worth it. Once you've had to replace your first clutch, you'll probably agree with me.

TURBO BOOST

In a turbocharged car, a boost gauge is a vital tool – both in seeing what's happening on a day-to-day-basis and also in assessing performance mods. Strangely enough, many people forget all about the latter use!

To use a boost gauge in this way, you'll need to correlate boost with another variable – usually engine rpm. Using an assistant (to drive or write), note down the boost reading each thousand rpm in the one gear. If the rpm rises too fast to do this easily, do it in a few goes – eg firstly 2000, 4000, 6000 rpm, and then, on another run, 3000, 5000 rpm – just as we did when using the accelerometer in the previous chapter. In this way you get a curve that can be graphed.

Shown in Figure 3-3 are four different boost curves recorded by my former colleague Michael Knowling. They show the boost behaviour of a turbo on a car that did not use electronic control of the boost level. The

If you own a car with forced induction, the readings from a boost gauge, when correlated with engine rpm, can give you lots of useful information.
(Courtesy Racetech)

orange line was the standard rate of boost increase, while the blue line showed what happened with the wastegate hose disconnected. The yellow and grey lines showed how Michael was able to tweak the boost control to bring up boost faster than standard, without overshoot.

This sort of modification is cheap and gives a real-world performance advantage.

Shown in Figure 3-4 is the performance of another very cheap boost control, this time working on a Nissan Skyline. I didn't want boost coming up any faster (ie I wanted factory throttle control) but I wanted the boost to rise to a higher level. As can be seen, the controller achieved its aims.

Another way of using a boost gauge is to detect when a certain boost level is first reached. This can of course be done with a full graph, but it's easier to just record the revs. When I was modifying a diesel Peugeot, one of my aims was to bring up boost more quickly. With a new exhaust, a level of 7psi of boost was achieved 800rpm earlier in the rev range (and that's a lot on a low-revving diesel!).

In addition, graphing the boost curve can reveal problems like:

Boost spikes – A boost spike is a sudden overshoot from the designated boost pressure. For

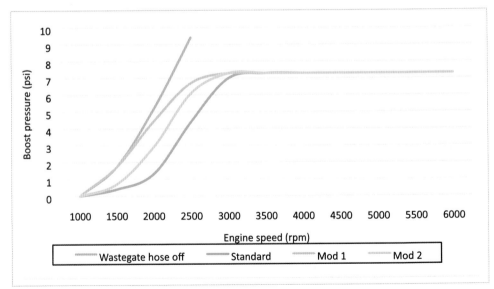

Figure 3-3: The measured boost curves using different approaches to wastegate control. The final configuration (Mod 1) brings up boost much more quickly without overshoot. Measuring boost on the road against engine rpm gives very useful information.

MEASURING ON-ROAD PERFORMANCE

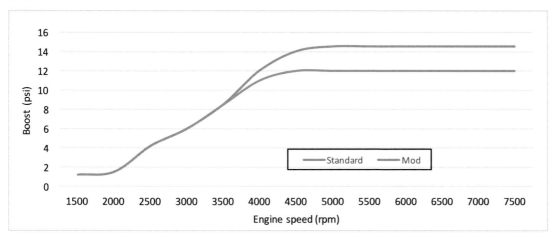

Figure 3-4: The intention of the modification was to give a standard rate of boost increase, but to peak at a higher level. The modification did just that.

example, if boost is set to 15psi but during hard acceleration rises for 1 second to 20psi, before again dropping back to 15psi, a spike has occurred. Boost spikes that occur when the engine is coming on boost are common in poorly set-up pneumatic systems, or electronic systems with insufficient damping.

Boost fall-off – This occurs when the boost level reaches a peak (eg at max torque) and then gradually falls in level through the rest of the rev range. This can occur in poorly set up systems that don't have any feedback (ie open loop systems).

Surging boost (full throttle) – If the boost at full throttle constantly varies up and down (eg by a few psi) the car is said to have a full-boost surge problem. This is common in poorly set up or badly designed electronic boost control systems featuring feedback that has insufficient damping.

If you have a turbo car, ensure that you use a boost gauge for more than just detecting maximum boost.

GEAR CHANGE POINTS

Besides plotting torque and power curves (see the previous chapter), another very effective use of an accelerometer is to work out the best gear change points. Instead of

Working out the best gear change points is best done by plotting the acceleration curves in each gear. This takes into account the gear ratios, engine torque curve, rotational inertia and aerodynamic drag.

acceleration being plotted against engine speed as was described above, it is plotted against road speed. The acceleration in each gear is measured, with the numbers noted each 5 or 10km/h (mph).

Each gear is tested from as low a speed as possible to the maximum speed possible before the engine redline. Doing this results in the sort of diagram shown in Figure 3-5, which is for a turbo 2-litre. As can be seen in this diagram, in this particular car, the maximum possible acceleration occurs if all gears are held to the redline. This is because each time a gear change is made, acceleration

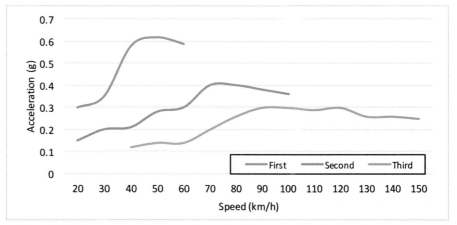

Figure 3-5: The measured on-road acceleration curves in the first three gears. As can be seen, maximum acceleration occurs if every gear is held to the redline.

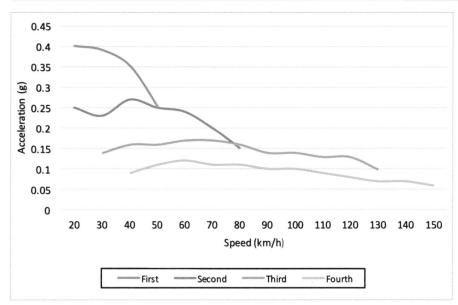

Figure 3-6: On this car you can see that best acceleration occurs if the second-third gear change occurs a little earlier than the redline.

drops markedly. The first gear change drops acceleration by a massive 47 per cent – implying that the jump in ratios between 1st and 2nd gears is too great.

However, as you can see in Figure 3-6, short-shifting will give quicker acceleration in some cars – this is for a naturally-aspirated 2-litre car. Here, a slightly early change from second to third gear will improve acceleration. This is because holding second gear to the redline results in the car being slower than if the early change to third gear is made.

In some cars, especially turbo diesels, performance can die away quickly at higher rpm. However, how much the performance drops depends on how much boost is being developed – and that in turn depends on the magnitude of the load and how long the engine is at that load. It becomes a bit of a 'chicken and the egg' scenario, but in these cars the shape of the torque curve is quite gear-dependent. This can result in the best revs at which to change gear varying substantially from gear to gear.

Incidentally, be wary of those who state categorically that 'of course' the best gear change points can be calculated, eg from a dyno graph. As I have said, sometimes the shape of the torque curve varies from gear to gear, and at given engine revs, aerodynamic drag certainly varies from gear to gear – both factors making the calculation much more involved than it first looks. In a modified diesel turbo I owned, best gear change points were: 1st-2nd at 38km/h (4750rpm), 2nd-3rd at 65km/h (4250rpm) and 3rd-4th at 94km/h (also 4250rpm). None of that could be worked out from the single-gear dyno graph that I had on hand!

TORQUE CONVERTERS AND AUTOMATIC TRANSMISSIONS

Most of the graphs shown in this book are for cars with manual transmissions. However, cars with traditional auto transmissions that use a torque converter have a significant performance advantage when accelerating off the line. That advantage is caused by the torque multiplication of the converter, which gives very strong standing-start acceleration. Figure 3-7 shows the measured acceleration of a 3.5-litre six-cylinder equipped with an automatic transmission, through

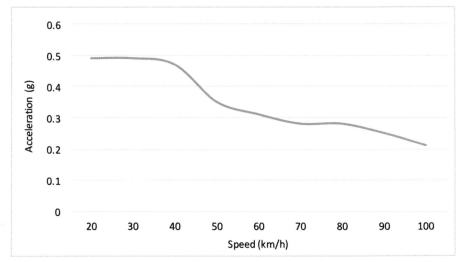

Figure 3-7: The on-road acceleration curve of an automatic transmission car that uses a torque converter, measured from a standing-start. Note the strength of acceleration off the line!

MEASURING ON-ROAD PERFORMANCE

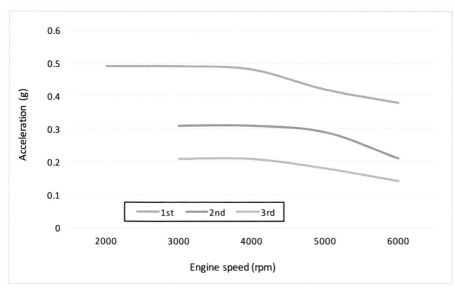

Figure 3-8: The same automatic transmission car as in Figure 3-7, but with each of the first three gears graphed. Note how it looks as if the drop in acceleration would be massive with each gear change, but because of the action of the torque converter, the change in ratios is, to an extent, masked.

first, second and third gears, starting from a standstill.

Note how in first gear (that takes the car to a road speed of 40km/h), the acceleration is constant and strong. The torque converter's multiplication, which is greatest when the wheels are stationary, is most clearly shown in this first gear. In second gear, from 40-80km/h, the acceleration starts to taper away as the relatively poor top-end breathing of this engine takes effect, and then in third gear (from 80km/h) the acceleration curve reflects the better bottom-end torque, trailing away as revs rise.

Figure 3-8 shows the acceleration curves for each gear of the same car, this time with each gear measured separately. Note how in this graph it looks as if the drop in acceleration would be massive with each gear change, but because of the action of the torque converter, the change in ratios is masked. However, looking at both Figures 3-7 and 3-8, it would appear that 2nd gear could be a bit shorter in ratio, so reducing the drop in acceleration experienced in this gear. Incidentally, if you have an easy way of altering engine torque output depending on the gear you're in (eg it's a turbo car, or you have programmable engine management), then you could achieve a flatter acceleration curve by juggling outputs on a gear-specific basis (for example, if this were a turbo car, by running slightly higher boost in 2nd gear).

INTAKE AIR TEMPERATURE

All engines develop more power if the intake air is colder and so more dense (ie it contains more oxygen per unit volume). Intake air temperature can be measured using an added sensor, by the engine management sensor output accessed via the OBD port data stream, or via the readout on a programmable engine management laptop. Note that here I am talking about the temperature of the air measured in the intake plenum, or close to it.

Intake air temperature is important in all cars but in those with forced aspiration (ie a turbo or supercharger) it becomes even more critical. This is because the compression of the air causes a rise in temperature (and so the air is more likely to be hot) and hot intake air helps create the conditions for

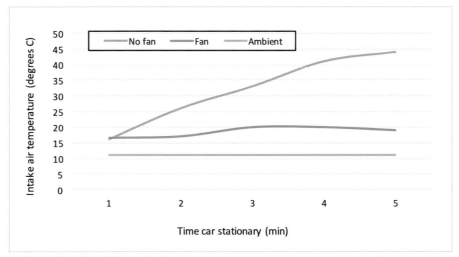

Figure 3-9: The different intake air temperatures recorded on a stationary turbo car, with and without a fan operating on the intercooler. Note the ambient temperature line.

engine-destroying detonation.

In cars driven mostly in urban conditions, it is when the car is stationary in traffic on a hot day where highest intake air temperatures are measured. As an example of using this data in the development of a modification, I added a fan to a turbo car intercooler. The intercooler was located in the engine bay and fed air by a bonnet (hood) scoop. With the car stationary, the scoop started to act as a chimney, with hot air flowing out through the intercooler and scoop. In this situation, the intercooler was acting as a heater, not a cooler! As a result, that car lacked power off the line and was also more susceptible to detonation in these conditions. Adding a fan that was triggered only when the car was stationary kept outside air flowing through the intercooler. Figure 3-9 shows the measured results.

Note that I have found dyno testing of intercoolers (and radiators, oil coolers, etc) to be quite at variance with the operation of such heat exchangers on the road.

Especially with complex intercooling systems such as this water/air design, measuring intake air temperature on the road gives a far better insight into what is actually happening than dyno testing.

Chapter 4
Flow testing intakes & exhausts

All combustion engines require the flow of gas. In a petrol or diesel engine, the flow of intake gas (air) starts at the intake to the airbox. The air flows into the airbox, passes through the filter, exits the airbox and then often passes through an airflow meter. In some cars it then moves straight onto the intake manifold; in others it passes through a throttle butterfly. There might also be one or more turbo compressors ahead of the engine.

If at any point in this process there are restrictions, the pressure of the air will decrease. Therefore, by measuring the pressure drop through the intake system, you can quickly identify whether the intake is posing undue restriction. Clearly, the process will work best if the airflow is at the same levels that occur in the engine when it is operating; therefore, the technique is most accurate if it is undertaken when the car is being driven on the road. You can also make changes and see the effect by, for example, temporarily removing the air filter and seeing the change in flow restriction.

On the exhaust side, the exhaust gases are forced out of the engine by the rising pistons. The gas then passes through a catalytic converter, perhaps a particulate trap or NOx adsorber, before then moving through one of more resonators and mufflers. If the exhaust flowed perfectly, there'd be no back-pressure. But we know that's not the case, and so by measuring how high the pressures actually are, we can see how well the exhaust is flowing. Once again, the process is most accurate if it is undertaken when the car is being driven on the road.

By measuring pressures in both the intake and exhaust, you can quickly and easily get a really accurate picture of what is occurring.

So how do you make these measurements? Let's start off with intake systems.

MEASURING INTAKE SYSTEMS

On the intake side, restrictions will show up as pressure readings that are *below atmospheric*. That is, if there's a restriction to flow, not all the air available from the atmosphere will 'get through' – resulting in a pressure drop. Measuring pressures below atmospheric can be done in a number of ways.

Cheapest is to make a simple water manometer – it will cost you nearly nothing, be highly accurate, and can be easily made. On the downside, using it can be unwieldy. Instead of the water manometer, you can use a Dwyer Magnehelic gauge. These gauges, designed primarily for use in air-conditioning and ventilation systems, are cheap secondhand, accurate, durable and convenient. I use these gauges. Thirdly, you can use an electronic digital manometer. This will cost you more (though still not a lot), but these instruments typically cover

SPEEDPRO SERIES

a wider measuring range and are convenient and quick to use.

Let's look at each in turn.

Water manometer

It's easy to make a manometer – here's how.

A water manometer made from a plastic soft drink bottle, some clear tube and a piece of wood. The free end of the hose (top) is run to the intake system before the throttle. The bottle is filled to the line marked 'water,' and then the car driven. The lower the pressure in the intake, the greater the restriction to flow and the higher the water rises in the tube. This piece of wood is marked off in centimetres of water, but you can use any units for length.

Get a large plastic soft drink bottle, and attach a thin vertical piece of wood to it with adhesive tape. Run a clear plastic PVC tube down the timber and place the bottom end of the tube in the soft drink bottle. Fill the bottle with water to a mark, and further place markings on the wood – 1 inch above the water level, 2 inches above, and so on. Run the free end of the tube to the point on the intake where you want to measure the pressure. If the pressure at the end of the tube is lower than atmospheric, the water will be drawn up the tube. A pressure of 1 inch of water below atmospheric will show as the liquid rising 1in up the tube. (2 inches of water below atmospheric, the water will rise 2 inches, etc.) You can add food colouring to the water to make it easier to see.

Magnehelic gauges

The Dwyer Magnehelic gauge was invented in 1953. The gauges are large (100mm/4in diameter), easy-to-read instruments that use diecast aluminium cases and plastic faces. Internally, the movement of a silicone rubber diaphragm is transmitted to a pointer without the use of gears or other direct mechanical linkages. This approach has some significant advantages over other gauge designs.

First, the use of a large diaphragm means that the pressure gauge can be much more sensitive than one using a traditional Bourden tube. Another reason that the sensitivity of the gauge can be so high is that the diaphragm movement is transmitted to the gauge pointer magnetically, avoiding physical contact that can also cause hysteresis (backlash) and jerkiness.

Magnehelic gauges are fairly expensive if bought new. However, partly because they have been

A Magnehelic gauge. This one goes to 150 inches of water which is a bit high unless your intake is incredibly restrictive. Better to select a gauge that has a maximum of 25 or 30 inches of water.

available for so long, and partly because many sellers don't seem to know what they have, these gauges are available quite cheaply on eBay. There are a few points to keep in mind when buying a gauge. Pick a gauge that has a range of 0-20 or 0-30 inches of water, then if you find yourself using this gauge a lot, go for a second more sensitive gauge (eg 0-10 inches) so you can see what's happening in more detail.

Note that most Magnehelic gauges have a legend on the face that says 'max pressure 15psig'. This means that the maximum pressure that the gauge can be subjected to as a whole is 15psi above atmospheric (this is not relevant in our testing). I mention this because many sellers will quote this as the gauge maximum, ignoring the much lower maximum reading that's actually on the scale. So if buying, it's best to be able to sight a photo of the scale so you can be sure of what you are getting. Note that the gauges are available in many different units, so don't be put off if the meter is marked in Pascals for example – just do an online conversion to inches of water to see

FLOW TESTING INTAKES & EXHAUSTS

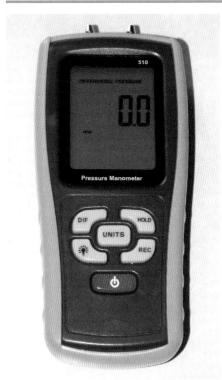

A digital manometer can also be used for measuring intake pressure drop, and so flow restriction. An advantage of using such a meter is that if one with the correct range is selected, the instrument can also be used for measuring aerodynamic pressures – see Chapter 6.

if the gauge is suitable for your use.

Digital manometers

Digital low-pressure electronic manometers are available, and while they're more expensive than a secondhand Magnehelic gauge, they also cover a wider range of pressures. Some of these instruments can also average readings, and have maximum and minimum peak-hold facilities that can be useful.

MAKING MEASUREMENTS

In a naturally-aspirated car, measuring the total pressure drop (ie total restriction) is as simple as using a quarter-inch rubber hose to connect the vacuum port of the digital manometer or Magnehelic gauge to the intake system just in front of the throttle butterfly, and then pulling peak power from the engine on the road. (With the water-and-tube manometer, the long length of free tube goes to the intake. Make sure that you don't suck the water out of the manometer – this might happen if the intake system is incredibly restrictive!) In a turbo car, the maximum intake restriction is measured in front of the turbo. In a naturally-aspirated car without a throttle butterfly, you can plumb the manometer or Magnehelic gauge to the intake manifold.

Connection to the intake system to measure total pressure drop can normally be made by temporarily pulling a breather hose and using that point to make the connection. The greater the restriction, the higher the low-pressure (vacuum) reading on the instrument or gauge.

TOTAL PRESSURE DROPS

The first time I ever did this sort of testing was on the intake system of a turbo Subaru Liberty RS, the predecessor to the famous Subaru Impreza WRX. With an assistant in the car and a self-built water manometer placed vertically on the passenger side floor, I raced up a steep local hill in second gear, foot flat to the floor. At that stage I didn't even know if the approach would work, let alone how high the water might go. My assistant called off the numbers as revs rose – 10 inches! ...15 inches! ... 20 inches! ... 25 inches! Her voice rose higher as she called the next figure – 30 inches!

And then, knowing that the manometer height extended only to 31 inches, I lifted throttle. As soon as the airflow into the engine decreased, so did the measured pressure drop – and the height of the liquid in the manometer plummeted.

Testing countless different cars since has shown that a maximum total pressure drop of over 30 inches of water in the intake system indicates plenty of restriction. Another car with a poor intake was the turbo Audi S4 I once owned. At full load, it had a maximum intake system pressure drop of 32 inches of water. And, as you'd therefore expect, modification to the intake systems of both of these cars resulted in good performance gains.

On the other hand, some cars have extremely good intake systems – eg, ultra-efficient cars like petrol-electric hybrids. The NHW10 Prius has a measured total pressure drop before the throttle body of just 10 inches of water (incidentally, the same as recorded on a Holden JE Camira), while a Gen I Honda Insight has a measured maximum of just four inches of water – the best standard car I have ever measured. A Skoda Roomster 1.9 TDi has a max of 14 inches of water, an EF Ford Falcon six-cylinder has a max of 16 inches of water, a Toyota Crown Supercharger 20 inches of water,

A miniature plastic irrigation fitting screwed into a hole drilled in the wall of an airbox. This fitting allows pressures to be measured at this point. When testing is complete, it can be unscrewed and a dob of black silicone placed over the hole.

SPEEDPRO SERIES

> ### FLOWS VERSUS PRESSURES – WARNING!
>
> Let's take a step back and think about what we are measuring. As I have said, the more restrictive the intake, the lower the pressure. However, there is another factor that will lower pressure as well, and that's how fast the air is flowing through the duct. The smaller the cross-sectional area of the duct, the faster will be the airflow and so the lower the pressure. In the real world of practical testing, this effect isn't a huge issue. However, to avoid it, it's best to do one or both of the following:
>
> 1) For greatest absolute accuracy, measure where the airspeed is slow. An example of this is on the engine side of the airfilter, in the airbox.
>
> 2) Make only comparative readings. That is, if the vacuum recorded at a location decreases after you have made an intake modification, and the cross-sectional area of the duct is unchanged, it's because the flow restriction has been reduced.

and a Peugeot 405 SRDT also 20 inches of water.

From these figures you can see total pressure drop of about 10-15 inches of water is a good achievable aim. In other words, if you measure a car and in standard form, it's at 10 inches of water, there's not going to be a lot of possible improvement to be easily made. On the other hand, if it's at 25 or 30 inches, you can be certain that major improvements in flow will be possible.

The next step is to work out exactly where in the intake system the changes should be made.

FINDING AND FIXING THE PROBLEMS

Within the constraints of the variation in pressures caused by differing airflow speeds (see the box above), the same pressure measuring technique can be used to find *which parts* of the intake system are most restrictive – and which are not.

So how do you find which parts of the system are most restrictive? The short answer is that you move the pressure sensing hose around, tapping into different points. In many cases, this will mean that you'll need to drill small holes in intake ducts, the walls of the filter box and so on. Screw a small fitting into the hole (I use miniature plastic irrigation hose connectors that self-tap into holes drilled in plastic) and then connect the sensing tube to the fitting. The holes need only be tiny, and are easily later concealed with a wipe over with black silicone.

So for example, starting at the throttle, you might want to make measurements just before the throttle, at the outlet duct from the airbox, on the outlet side of the filter, on the intake side of the filter, and at the entrance snorkel to the airbox. (In a car with an airflow meter, you'd add two measurements, one either side of the airflow meter.) Now all this might sound like a lot of work, but it can all be done in literally 30 minutes of on-road testing. You need to use full throttle and run right to the redline, and I usually do the measurements in second gear. After you've done the measurements, you'll end up with a table like the one shown below.

(I'll use 'inches of water' as the units of vacuum here, but there are other units that can be used, of course.)

The table below shows the readings taken on a six-cylinder car that uses MAP sensing to determine engine load (so there's no airflow meter). It's an older car, but the design of intakes has not changed much in decades. The total pressure drop of the standard intake system is shown as 16 inches of water, so not too bad at all – helped by the lack of an airflow meter. Of that total, the air filter is causing only 1 inch of water restriction. The greatest flow restriction is occurring at the outlet to the airbox – it makes up nearly 38 per cent of the total restriction.

Figure 4-1 shows the pressure drops in graphical form. As you can see, in the case of this car, the priorities were to improve the duct between the airbox and the throttle,

Location	Total pressure drop	Difference	Description of section of intake
Before throttle	16.0		
		2.8	Dual duct between box and throttle
Outlet duct of airbox	13.2		
		6.0	Airbox outlet
Outlet side of filter in airbox	7.2		
		1.0	Filter
Intake side of filter in airbox	6.2		
		3.0	Snorkel
Intake of snorkel	3.2		

FLOW TESTING INTAKES & EXHAUSTS

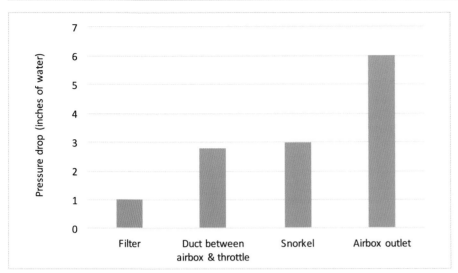

Figure 4-1: The pressure drops across each part of the intake system, measured at full load on the road. Note how little restriction the standard filter causes. In all of the cars I've tested, the filter has never been the major restriction.

the airbox outlet and the intake snorkel. The photos below and on the next page show how I went about doing so.

And what was the result of the simple and cheap modifications? With the modified intake system, the full power pressure drop was reduced to 9 inches of water – an improvement of 44 per cent! On the road the engine revved out more freely above 4000 rpm, and the throttle response at all revs was clearly sharper.

In the case of the testing described above, I chose to show just the peak pressure drops – that is, those recorded at maximum engine power. However, just as we did when measuring instantaneous acceleration in Chapter 2, we can also plot the intake pressure drops against engine rpm. Figure 4-2 on page 33 shows this done for the total intake pressure drop of an Audi S4. Note how the shape of the curve is very much like a power curve – power is proportional to the mass of air the engine is breathing, so here is another way of seeing the power curve!

EXHAUST SYSTEMS

Much the same approach is taken when measuring the restrictiveness of the exhaust system. However, in this case, a normal positive pressure gauge can be used – eg an old turbo boost gauge.

The tapping points into the system can be achieved in two different ways.

The standard 4-litre, six-cylinder engine's intake system comprises a snorkel that picks up air from between the bonnet (hood) and its locking platform, a large airbox, and a twinned duct that goes from the airbox to the throttle body. Maximum pressure drop before the throttle was 16 inches of water – not too bad, but able to be improved. It's an older engine, but little has changed.

The duct between the airbox and throttle was replaced with the larger intake duct from a later model car. Cheap, simple and easy!

SPEEDPRO SERIES

The airbox outlet was replaced with this much larger duct. The new duct was made from 75mm (3in) plastic stormwater pipe. The end was heated over a hot plate, and then forced down over a small upturned bowl, flaring it into a bellmouth. Using a sanding roll held in an electric drill, the hole through the airbox was enlarged slightly until the new pipe was a tight push-fit. The pipe was cut to length to provide a good match with the new intake duct, and then the PVC pipe was sanded smooth with wet-and-dry, and painted black with a spray can. Cheap and easy!

The standard intake snorkel to the airbox was replaced with another factory Ford part. The intake mouth is much bigger than stock and the body of the snorkel also is a little larger. The factory rubber bonnet seal fits nicely around the new snorkel which screws straight into place.

But what about the small bonnet (hood) gap that fed the airbox intake snorkel? The easiest way of overcoming that was to connect the airbox intake snorkel to a high pressure area – like the one existing in front of the radiator. This new flow path was easily achieved by installing a new 75mm (3in) diameter plastic pipe from adjacent to the mouth of the intake snorkel down through two plastic panels to the area in front of the radiator. Heating and bending the plastic pipe gave a nice finish – easily achieved with a heat gun or electric stove and some simple hand tools.

FLOW TESTING INTAKES & EXHAUSTS

A forward-facing elbow was placed at the base of this duct, with the mouth flared with the heat-and-bend technique. The elbow is optional – just connecting the pipe to the area in front of the radiator will allow it to pick-up high-pressure air.

If you want to measure the total back-pressure, the air/fuel ratio sensor (oxygen sensor) can be temporarily removed, and a replacement fitting screwed-in in its place. Run a hose to the fitting and measure the total back-pressure at peak load. (If the testing is going to be extensive, you will want to make the first section off the fitting a metal pipe, allowing it to shed some heat before the hose – otherwise the hose can melt.) Because no gas actually travels inside the hose to the gauge, the gauge won't get hot and be damaged.

If you want to measure the contributions that the individual parts of the exhaust make to the total, you'll need to drill holes at different points in the exhaust pipe and move the sensing hose. Unlike the intake system, where these holes are later easily filled, the exhaust will need to be welded to seal the holes. Therefore, you'll be pretty well locked into having work done on the exhaust after doing this testing. (So that's why it makes sense to do the total back-pressure measurement first!) The holes can be temporarily sealed by the use of a hose clamp around the tube.

Exhaust back-pressure should be measured after the turbo, or after the exhaust manifold/factory headers.

So, what sort of figures are you likely to see when back-pressure testing exhausts? On standard cars I've seen readings as high as 13psi, and many cars are running at 8-9psi. Looking at individual sections of the exhaust, on one slightly modified turbo car, the section of pipe off the

THROTTLE BODIES

So, how can you measure the flow of throttle bodies? If you connect a pressure measuring tube to the engine side of the throttle, the measuring device will see full engine vacuum when you lift off – usually not good with sensitive instruments.

To test throttle bodies, I suggest you use a relative pressure measurement – that is, you run tubes to each side of the throttle and connect them to the appropriate high- and low-pressure ports of the Magnehelic gauge or digital manometer. The low-pressure port will connect to the engine side of the throttle, and in this tube place an on/off ball valve. (A ball valve goes from fully open to full closed with just a half turn.) Working with an assistant, have them shut off this ball valve until you have reached full throttle (preferable in a high gear). At full throttle, high rpm, get them to open the valve, read the value being displayed and then shut the valve before you back off.

This way, you should be able to accurately measure the flow restriction across just the fully-open butterfly valve.

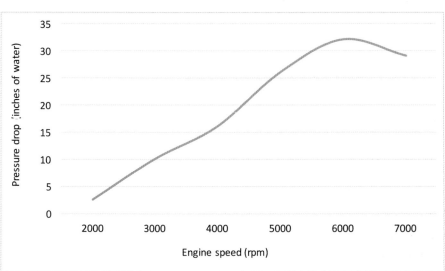

Figure 4-2: Total pressure drop of an intake system, plotted against engine rpm. You can see that the pressure drop curve replicates the power curve of the engine. This is for an Audi S4 – in standard form, peak pressure drop was a high 31 inches of water.

SPEEDPRO SERIES

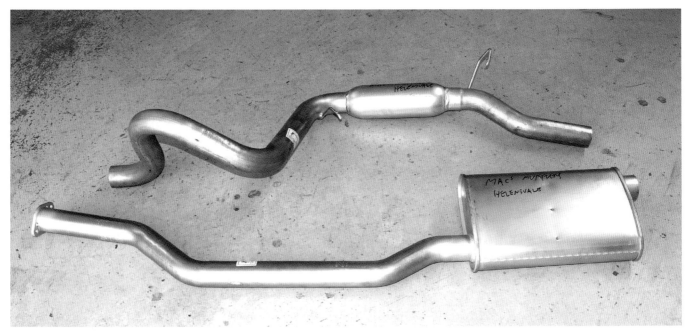

A new aftermarket exhaust, about to go onto a car. But is it actually needed? A simple on-road test will find out how well the standard exhaust flows.

turbo was causing 25 per cent of the total, the catalytic converter 19 per cent, the pipe between the cat and the rear muffler 26 per cent, and the aftermarket reverse flow rear muffler 30 per cent.

FIXING PROBLEMS
All the following approaches to improving exhaust flow are valid:
- Better dump pipes off turbos
- Free-flow mufflers
- Larger cat converters
- Larger diameter (or twinned) pipework
- Mandrel bends.

In most cases, the cat converter is the most restrictive element, followed by the main muffler.

Unless bends are very large in angle (eg 180° or more), press-bending does not normally cause much measurable restriction. Going up one size of tube and using press-bends is usually more cost-effective than buying mandrel

A high-flow exhaust made from salvaged components, including a GM catalytic converter, Jaguar straight-through muffler and Ferrari exhaust butterfly! Fitted to a turbo car, I later had to add a central resonator – and I bought that new. This exhaust flows very well.

FLOW TESTING INTAKES & EXHAUSTS

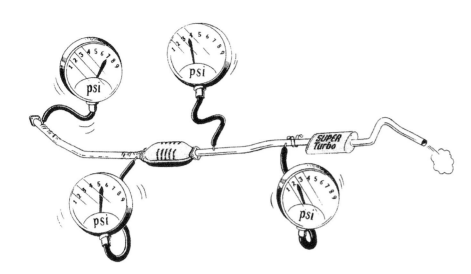

By measuring the back-pressure at different points along the exhaust, the restriction of each part can be found. Here total back-pressure is 7psi, with a substantial amount of the flow restriction occurring across the catalytic converter. (Courtesy Dave Heinrich)

That will immediately show you the potential for improvement – whether there's a lot or a little. If there's a lot, you can then do some more measurements to find out which parts of the system are most restrictive. Again, at absolute minimum, measure the total exhaust back-pressure before spending anything on modifying it.

Measuring pressures through intake and exhaust systems is simple, cheap and extraordinarily effective. It adds a huge flowbench to your tool kit and simultaneously takes into account all the real-world conditions that occur.

The flow restriction of different muffler designs is easily measured by on-car testing. This muffler (sometimes used as a resonator) uses flush holes within it, posing little flow restriction. Compare that design approach with punched louvres that project into the flow!

bends. Don't forget that if your car is not very powerful, fantastic bargains can be had by fitting the components from discarded standard exhaust systems taken from more powerful cars.

Typically, a good aftermarket exhaust (including one made up of secondhand items pinched from more powerful cars) will reduce measured exhaust back-pressure by around 75 per cent. To put numbers to that, a good exhaust (that includes a catalytic converter) has about 2-3psi back-pressure at full power.

CONCLUSION

At absolute minimum, before you spend anything at all on modifying the intake system, measure its total pressure drop at full load.

Chapter 5
Testing suspension & brakes

In this chapter I want to talk about testing vehicle suspension – that's ride and handling – and brakes. However, in many respects some parts of this chapter are rather different to the others. Why? Well, the greater the pressure drop through the intake system, or the slower the turbo boost is to rise, the worse will be the performance – no question about it. But *my* 'good handling' car may not be *your* good handling car. *My* 'excellent steering response' may be *your* twitchy steering. *My* 'excellent ride' may be *your* mushy, under-damped and wallowing barge!

While suspension can be measured with numbers (and we'll cover some such approaches shortly), often, it is best if it is evaluated *subjectively* – both the car's standard handling and after it's modified. In short, the primary tool-kit becomes the descriptive terms that you apply, rather than the numbers you measure.

Testing of handling using a low-speed slalom at a company proving ground. (Courtesy Holden)

Let's take a look …

HANDLING TERMS
- **Understeer**

Understeer is when the cornering car does not turn as sharply as the amount of steering lock indicates it should. In other words, the front of the car is sliding outwards. A car that understeers off the road does so with the front still facing the original direction. Understeer

TESTING SUSPENSION & BRAKES

is safer than oversteer and so all car manufacturers set up their cars to understeer when grip levels are exceeded.

A car with *plough understeer* has completely stopped responding to steering inputs – turning the wheel further makes no difference to the direction the car is headed. Plough understeer is dangerous because steering control has been lost.

Power understeer means that the understeer increases when more power is applied. This normally applies to front-wheel drive cars. Power understeer can be remedied by reducing power, so transferring weight forwards and also reducing the work the front tyres need to do.

Turn-in understeer is when the car is slow to respond to the steering when it is first applied. High-speed turn-in understeer can be very disconcerting because it feels as though you will not make the corner – and it doesn't feel like there is much you can do about it!

- **Oversteer**

Oversteer is where the car turns to a greater degree than the steering indicates it should. In other words, the rear of the car is sliding outwards. A car that oversteers off the road is spinning, so it may hit sideways or tail first (or even do a 360-degree spin, hitting nose first!). An oversteering car can be felt to be rotating around you.

A car with *power oversteer* has increasing oversteer with the application of more power. This normally applies to rear-wheel drive cars. Power oversteer can be remedied by gently reducing power.

Lift-off oversteer occurs when the throttle is abruptly raised mid-corner. This normally applies to front-wheel drive cars but will occur in any car with high rear roll stiffness. A car that lift-off oversteers will normally tuck-in if the throttle lift isn't so great – that is, the front will stop understeering.

Turn-in oversteer is when the car initially turns-in more than the steering angle requested. This is very disconcerting because, as with turn-in understeer, it's not immediately apparent what can be done to stop it.

- **Bump steer**

Bump steer occurs when the wheels change their toe angles (the direction they're pointed in) as the suspension moves up and down. It's generally most easily felt on turn-in, when a front suspension with toe-in on bump will have turn-in oversteer (it will twitchily turn-in more than expected) and a front suspension with toe-out on bump will have turn-in understeer.

- **Roll linearity**

A car with linear roll is progressive in its roll, with the angle of roll being directly related to how hard the car is cornering. A car with non-linear roll may rapidly lean on turn-in but then not lean any further as the cornering load increases. Roll linearity is seldom mentioned but it is very important in giving the driver the correct signals as to what is occurring.

- **Twitchiness**

This describes how rapidly the car responds to driver inputs.

A twitchy car will respond very rapidly to steering and power inputs. In a road car this tends to be tiring and at times disconcerting. However, a car that is the extreme opposite to 'twitchy' will feel dull and unresponsive. Stiff bump damping and stiff sway bars are two characteristics that will make a car twitchy.

- **Throttle steer**

A car that is on the edge of sliding (either understeer or oversteer) and can then be moved into a slightly sliding state by a variation in power is said to be being throttle-steered. Throttle steering requires a very well balanced car with an engine that has a linear torque response.

ANALYSING HANDLING

The above terms give you the vocabulary – but how do you apply it?

Firstly, to initially explore the handling of a car, the car does not need to be driven fast. One of the first things I do in a car new to me is to drive down a straight road, making quick but minor left/right/left/right steering inputs. Does the vehicle respond quickly or slowly? Is the steering around centre linear, or does little happen – and then as the wheel is turned further, suddenly a lot? Is vehicle roll in proportion to the

One of the first things I do when assessing handling is to drive down a straight road, making quick but minor left/right/left/right steering inputs. This sort of test can be done anywhere.

sideways forces (ie how good is the roll linearity)?

The next step is to apply power while cornering – and explore the areas of power oversteer, power understeer, plough understeer and so on. The slipperier the surface, the lower the speeds (ie safer) at which this exploration can be conducted. Finally, higher speed characteristics (eg turn-in oversteer), twitchiness and bump steer can be tested.

This sort of evaluation, done carefully and thoughtfully, can tell you an immense amount about a car's handling abilities. It obviously helps if you can recognise specific shortcomings (eg bump steer) – and that can be hard to do when you haven't previously experienced it. In a way, analysing handling in this methodical, step-by-step way is similar to measuring pressure drops through the intake system, as I described earlier. One modification approach (a very common one!) is to just look at the intake system and then throw at it a new filter and cold air intake. A better approach is to measure where the problems actually are and then fix just them. In handling, you can buy new springs, dampers and anti-roll bars – or you can carefully consider what the problems are and fix just those.

SKID PAN TESTING

Subjective evaluations are critical, but there is one area of car handling where numbers can work very well – skid-pan testing. So what is a skidpan?

Throw out any ideas you might have of drift merchants circulating a piece of bitumen in a lurid tail-out slide. Instead, picture a skidpan as simply being a marked circle around which the vehicle is quickly driven. Rather than a place to do slides, think of it as a corner that never ends.

RACETRACKS?

I am not a fan of testing road car handling on racetracks. On the racetrack, cars never come the other way towards you. The track is smooth – far smoother than most roads. The corners are known to you – they never unexpectedly tighten. Racetracks have run-off areas, and they don't have children suddenly running out in front of your car. You don't drive on race tracks when you're tired, stressed – and then all of a sudden, there are roadworks and it starts to rain. In my experience, a car with handling that is quick, nimble and fast on a track is twitchy, tiring and 'sudden' on the road. And I don't want any of those things in my road cars ...

The 'continuous corner' idea immediately highlights the importance of a skidpan. Instead of being able to use a traditional cornering 'slow-in, fast-out' approach, you must maintain the cornering line. Rather than following a 'racing line,' the line you follow is prescribed. Rather than being able to get a higher cornering power for just an instant, it must be held continuously.

And to negotiate the never-ending corner, the steering must be accurate and sensitive, the motive power must be able to generate enough power to keep you going around as fast as possible, the tyres must continue to develop adequate grip, and the oversteer/understeer balance must be good enough to allow the driver to keep all the wheels pointing as much as possible in the direction of the corner. Because, if any of these things cannot be achieved, you'll be going slower!

And the aim of a skidpan is to go as fast as possible around the marked circle of a known diameter.

If you know the diameter of the circle and how long it takes the vehicle to get around it, you can then work out the vehicle's maximum lateral grip ('cornering g's). This number represents the fastest that a vehicle can go around a continuous corner, or to put it another way, the maximum sideways grip it can develop – usually, on a smooth, dry surface. Being able to directly measure the maximum sideways grip is a fantastically easy way of finding out something that is otherwise very hard to accurately calculate.

On a skidpan the fundamental steady-state handling traits can be explored – for example, the understeer/oversteer balance. In addition, the response of the vehicle to power changes is very easily determined – for example, whether the car throttle-lift oversteers. It is much easier to feel these aspects when continuously cornering than it is when only ducking in and out of corners.

I use skidpan testing a lot – and I do it in two quite different ways.

The first testing place could perhaps be called 'informal' – whenever I have a new car to test, I make sure that at some point I drive it very hard around a medium/large sized empty roundabout. I enter the roundabout relatively slowly and then accelerate to the point at which the car is just sliding. In every current unmodified car the front will lose grip first – it will understeer. I then lift the throttle fairly quickly and see what happens when the weight transfers forward. In many front-wheel drive cars, the front will tuck-in, and in some the rear will slide into

TESTING SUSPENSION & BRAKES

oversteer. In rear-wheel drive cars, the application of more power will cancel the front understeer and push the car into oversteer.

Clearly, sliding cars around public road roundabouts is quite politically incorrect – and may in fact be illegal in some jurisdictions. Any use by you of the technique is completely at your own risk – and also note that it requires far more skill than is shown by many drivers. However, as stated above, I am emphatically not talking about drift-style slides; instead, an observer would probably not even realise the car was sliding. If the steering wheel of the understeering car is turned so far that it is ploughing, you have not been listening to the car. If the steering is turned so far in opposite lock that it can even be noticed from the outside, you have let things go way too far. The beauty of an increasing speed skidpan test is that things happen relatively slowly and with plenty of warning; as a result, the corrections of steering and throttle can be achieved with subtlety.

Exactly the same test procedure applies to cars with electronic stability control; in fact, this is a really good way of feeling its characteristics. Some manufacturers allow quite a lot of driver control of a sliding car, while others shut things down very quickly.

I also suggest that a roundabout skidpan test is by far the safest way of road assessing a car that has had handling modifications. Especially when making mods that result in lift-off oversteer (for example, stiffer rear springs and/or rear anti-roll bar), it's vital to feel the car's behaviour when the car is on the edge of sliding. In these types of testing, the time taken to negotiate the skidpan circle is of far less importance that using the test to feel the car's characteristics.

If you're in a position where you can access some flat, empty bitumen (eg a large shopping centre car park when the shop is closed – get permission first!) you can mark out your own skidpan with some chalk. To do this, select the radius of the skid pan and cut a piece of rope to this length. Drive a nail into the bitumen to hold one end of the rope in place, and with the rope stretched taut, walk the other end around the circle, marking it with chalk.

The equation to work out the maximum lateral acceleration is:

$$\frac{39.48 \times radius}{time\ squared}$$

where radius is in metres, time is in seconds to complete the circle from a full-speed rolling start, and the answer is in metres per second per second. Divide this by 9.81 to get the results in g units.

The enormous skidpan at the Mercedes development track. Not all skidpans have to be as big as this to be effective for testing handing! (Courtesy Mercedes)

SKIDPAN DOWNSIDES

Of course, there are many aspects of grip and handling where a skidpan is useless. Turn-in handling behaviour cannot be assessed, transient lateral weight transfers are never judged, and it would be a fatal mistake to assume that a vehicle with a higher lateral 'g' figure is always going to out-handle one with a lower figure.

SPEEDPRO SERIES

SUBJECTIVE RIDE ANALYSIS

The other side of the coin to handling is ride quality. Again, while it is easy to measure data such as vertical acceleration (and only a little less easy to measures rates of change of that vertical acceleration), in the real world, a subjective analysis is often more useful.

Three aspects of ride can be easily judged.

- **Impact harshness**

This refers to the hardness of the ride when small, abrupt bumps are met – eg a sharp-edged pothole, or a rock lying on the road. The impact hardness that is felt is largely dependent on three aspects of the suspension – the tyre (profile and pressure), the spring stiffness and high-speed bump damping. (The latter two are virtually indistinguishable to the occupant of the car.) The effect of the tyres is easily separated from the springs and dampers – the ride feels like the tyre pressures are too high, even when they are not.

- **Low-speed damping**

'Low-speed damping' refers to the damper shaft speed, rather than the car speed. Low speed damping is the sort felt after the car has passed over a wave in the road. Does the car body continue to bounce or is the vertical motion quickly quelled? Or, is the motion stopped so fast that the ride feels too firm?

- **Bump vs rebound damping**

All vehicles use firmer rebound than bump damping. This is because, unlike bump damping, rebound damping does not make the impact of bumps harsher. However, a car can have too firm a rebound damping. The effect of this is hard to describe, but it can be clearly felt. What happens is that rather than being fast accelerated upwards over bumps, you feel yourself being pulled down harder. In other words, the bumps feel oddly reversed. A car with much too high rebound damping will also ratchet itself downwards on the springs if successive bumps are quickly met.

MEASURING SUSPENSION NATURAL FREQUENCIES

Let's say you want to compare the spring rates you are running in your car with someone else's. You're trying to sort out your suspension, and you are admiring a similar car where its handling and ride are excellent. You say to the owner: "What rate springs are you running?" He tells you what they are in pounds/inch and then adds: "But my car's motion ratio is different to yours – and the motion ratio on my car actually varies through the wheel travel."

The motion ratio is how much the spring compresses for a given movement of the wheel. So you think: hmm, that means that the rate at the wheels is going to be (1) different to the rate at the spring, and (2) at any given suspension deflection, is likely to have a different wheel rate to your car.

And then it gets worse.

The owner of the other car then adds: "And of course, my car's weight distribution is different to yours, what with this heavier engine and its slightly different location in the wheelbase. And don't forget, I don't care much about ride quality,

A stiff spring doesn't necessarily mean stiff suspension – it depends on the motion ratio and the weight acting through the spring. Measuring the natural frequency of the suspension takes away this confusion.

TESTING SUSPENSION & BRAKES

so my car's springs are probably stiffer than you'll want."

You think: gosh, and I thought comparing spring rates would be easy!

Well there is an approach that allows you to compare the suspension rates between different cars. It's a technique that takes into account motion ratio, spring rate, and mass acting through the wheels. It's called the suspension's 'natural frequency' and it allows direct comparison of suspension stiffness of different cars. It's why, in all good suspension textbooks, spring rates are never referred to – just natural frequencies.

So what is all this about – and then when we've got that sorted, how do we measure natural frequencies?

Let's say we take a coil spring out of a car and mount it vertically on a surface. We then put a heavy weight on top of it and push down firmly on the weight. When released, the weight bounces up and down at (say) three times per second – so at a frequency of 3 Hertz (Hz). Now it doesn't matter how far we push the weight down before releasing it, this combination of spring and weight 'likes' bouncing at 3Hz. This is called the system's *natural* or *resonant frequency*.

If we were to keep the spring the same, but change the weight, the resonant frequency of the system would change. It would also change if we kept the mass the same but altered the spring characteristics. To put this another way, there is a direction relationship between mass (acting downwards through the spring in this case), spring stiffness and the resulting resonant frequency. So if we directly measure the resonant frequency of the suspension, we get a number that takes into account the spring stiffness, the motion ratio of the suspension, and the mass that is working through the spring. The higher the natural frequency, the stiffer the suspension.

Sound hard to measure? It used to be, in fact needing lots of expensive gear. But these days, it takes less than a few minutes – you just use a cheap app and an iPhone or iPad. The app is produced by Diffraction Limited Design (see http://www.dld-llc.com/Diffraction_Limited_Design_LLC/Vibration.html) and is called Vibration. At the time of writing, it costs just US$6. (I'm told that similar apps also exist for other smartphone operating systems.)

The software takes advantage of the fact that the iPhone has an inbuilt 3-axis accelerometer. It can measure up to plus/minus 2.0g and has a sensitivity of about 0.02g. Those characteristics make it ideal for measuring suspension behaviour. The software can be set to sample at up to 100Hz (100 times a second) and the data can be displayed as graphs, or emailed as spreadsheets.

The first step is to download and then have a play with the Vibration app to see how it works. It's pretty straightforward, but like a lot of things, much quicker to learn by exploring the software functions on the phone than through my writing about it here.

Set the logging so it occurs for 10 seconds at the highest sampling rate possible – 100Hz. You can also set the sensitivity to suit the accelerations – when statically bouncing the car (covered in a moment), start off with 0.2g per vertical division. Finally, you can put in a delay that will occur prior to sampling starting – this allows you to get the car bouncing well before the logging actually begins.

After you have the functionality sorted, place the phone on one end of the car – say across the front axle line. Bounce that end of the car up and down. The suspension will strongly resist being bounced at anything but its natural frequency, so you very soon get a feel for when to push. (This is just like with a child's swing – it's obvious when to do the pushing.) With a car having stiff damping and/or spring rates, you might need a helper.

Press the sample button on the software and start logging as the car is being bounced. You should end up with an up/down trace that looks something like Figure 5-1. (It is the bottom yellow trace that shows vertical accelerations). This is the front suspension of a Honda Legend.

Switch the app to 'Frequency'

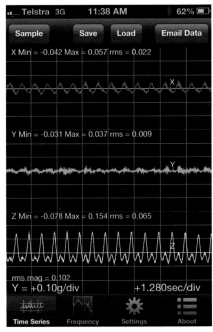

Figure 5-1: The measured front suspension behaviour of a Honda Legend using the Vibration app and a smartphone. The bottom trace, that shows up/down movements, is the one we're interested in. The next step is to switch the software to 'frequency' and do an analysis of this waveform to find the single dominant frequency.

and place the cursor on the peak. If there are many peaks, look for the one in the range of 1-2.5Hz – that will be the suspension frequency. Make a note of the reading. You can then do the same at the other end of the car.

In the case of the Honda, the front suspension frequency was 1.4Hz and the rear was 1.8Hz. The higher the resonant frequency, the stiffer is the suspension. In order to reduce pitch, most – but not all – cars have a higher rear than front frequency.

Below are the results of testing a mid-Eighties W123 Mercedes 230, one equipped with hydraulic self-levelling suspension at the rear.
- Front: 1.3Hz
- Rear: 1.3Hz

Note that overall, the suspension of the Mercedes is quite a lot softer than that of the Honda.

There's another thing to note as well. To be most representative of reality, during static testing, the car should be loaded as it normally is – eg with one or two people. Note that the Honda and Mercedes mentioned above were statically tested while unloaded.

Now, if you're thinking to yourself 'Couldn't the testing just be done by driving up and down the road?' – you're right. When testing on the road, the 'forcing frequencies' that road bumps impart are much more complex and varied than is achieved by simple bouncing of the car, but you can still normally identify the peak on which to place the measuring cursor.

In addition to testing for bounce frequencies, this sort of testing can be easily carried out for roll frequencies (ie how stiff the car is in roll) and pitch frequencies (how stiff the car is in pitching – where when the back is up, the front is down, and vice versa). These are easiest to do statically, by pushing on the car (to do pitch testing, you'll need two people, one at each end of the car, working in a co-ordinated manner). The roll frequency will be typically higher than the bounce frequencies – that's because of the additional spring of the anti-roll bars. However, that isn't the case with the Honda Legend:
- Front: 1.4Hz
- Rear: 1.8Hz
- Roll: 1.8Hz

The roll frequency of the Honda is the same as the rear frequency. This is presumably the case because the front springs are being stiffened enough by the anti-roll bar, that in roll they are as stiff as the rear springs. (And the rear anti-roll bar is very soft – which it appears to be.) The pitch frequency of the Honda is 1.6Hz – as you'd expect, numerically between the front and rear frequencies.

Figure 5-2 (below) shows the on-road measured data for an E500 Mercedes with air suspension. Again, we're interested in the lowest trace (circled). Note how the on-road data is more complex in shape than that measured when statically bouncing the car. However, a frequency analysis can still be performed (Figure 5-3). This shows that the natural frequency of the Mercedes suspension is 1.9Hz.

In addition to comparing the effective suspension rates of

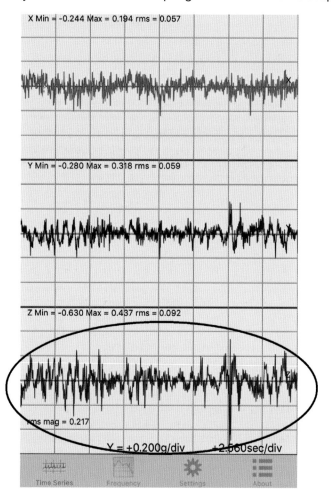

Figure 5-2: The recorded accelerations of a Mercedes E500 equipped with OE air suspension, as measured on the road in Sports 1 suspension mode. Again, we're interested in the bottom trace, that shows vertical accelerations.

TESTING SUSPENSION & BRAKES

different cars, what other use is this data? Most important is when you are making modifications. For example, if the rear springs are softer than the front, almost everyone will assume the rear suspension is also softer than the front. But that isn't necessarily the case – if the rear weight is less than the front weight, the rear natural suspension frequency could be higher than the front! So when selecting new springs, ensure you know what the existing actual suspension stiffnesses are – otherwise you can start heading off in quite the wrong direction. (This isn't a book on suspension modification, but for example in a front-wheel drive car, a car with existing high rear suspension stiffness that is made even stiffer can become dangerously unstable with quick changes of direction at high speed.)

Another really important use for this data is when you can easily alter spring stiffness, as you can on a car with air springs. (This is achieved by plumbing extra volumes in series with the air springs. The bigger the connected volume, the softer the spring rate.) In order to assess what the extra volume is doing, you can do a quick natural suspension frequency measurement.

Finally, in cars with variable rate springs, working out what the actual spring stiffness is (eg in lb/in) becomes quite problematic – it extends over a range and varies with compression. However, it's easy to measure their effective stiffness with an in-car natural frequency measurement.

TESTING BRAKES

Think of testing brakes, and most people will think of stopping distance – when the car is braked hard from a certain speed, how far does it travel before coming to halt? However, here is something you may not have thought of. If the car is capable of locking (or more precisely, near-locking) the brakes, the stopping distance does not actually depend on the brakes – it depends on the ABS logic, the tyres, the friction of the road surface, and how well the tyres' contact can be maintained with the road. The latter is in turn highly dependent on the dampers. Therefore, when we are talking about brake performance, it's in fact usually the *resistance to fade*, and *brake feel* that we're most concerned about, rather than stopping distances as such.

Resistance to fade can be measured using an accelerometer and by performing multiple maximum-braking runs in quick succession. This type of testing was done on an ABS-equipped family sedan, using a Lev-O-Gage accelerometer and testing on newly-laid bitumen. Ten consecutive stops were carried out from 100km/h (~60mph), with less than a minute between each stop. As with most such tests, peak deceleration (1g) was achieved just prior to stopping, however in the middle zone of braking, readings varied from 0.9 to 0.97g, with the first test (with the brakes still cold) showing 0.87g. Therefore, in this car, no fade was detected in this test – retardation remained strong and consistent. Interestingly, on a dirt surface,

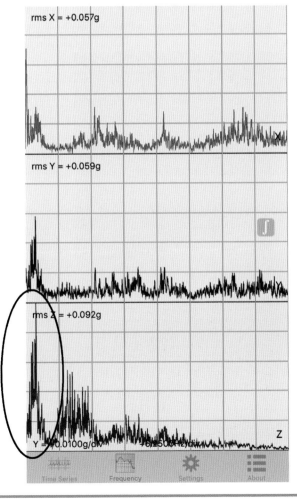

Figure 5-3: The screen grab shows the frequency analysis of the Mercedes E500 vertical accelerations. The ellipse has been placed around the low frequency spike, indicative of the natural frequency of the suspension. Note the single big spike, which is at 1.9Hz.

SPEEDPRO SERIES

maximum deceleration dropped to just 0.47g!

Some testing was carried out on a BMW 318i, one with noticeably worn standard suspension. The testing was of stopping distances from two speeds – 60km/h (about 35mph) and 80km/h (about 50mph). The suspension was then upgraded with new aftermarket sports suspension parts, including new springs and dampers, and the tests done again. The results are shown in the table below. The stopping distances were radically improved as a result of the tyres' better contact with the road.

However, in assessing brakes, I am more of a fan of on-road driving tests. When I last fitted big brakes to one of my cars, I wrote the following report after they had been fitted:

The ability of the brakes to pull the car down from high speed with progression and feel is superb – vastly better than with the standard brakes. On the road you can drop from 150 to 60km/h (about 90 to 35mph) with literally a gentle push of the centre pedal – and do it corner after corner. In an emergency stop, the ABS operates as it did with the standard brakes.

Downsides? There are some.

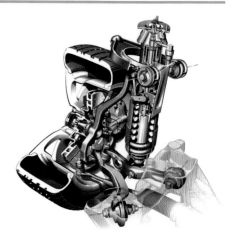

The performance of the brakes is heavily dependent on both tyres and the suspension, so ensure these are both in good condition before upgrading brakes. (Courtesy Mercedes)

When the brakes are dead-cold (in my normal daily drive I arrive at a roundabout after a long distance of country driving that has had literally no brake applications at all) the pedal needs a distinctively firmer push. In normal urban braking, this greater pedal effort doesn't occur (the brakes must retain some heat) and in spirited driving, the pedal effort is clearly lower than with the standard brakes.

	Standard worn suspension		**New sports suspension**	
60km/h	1.75 sec	7.13m	1.41 sec	6.71m
80km/h	2.47 sec	16.6m	2.25 sec	15.4m

Testing of brakes is most easily done on the road using the criteria of 'fade' and 'feel' rather than by using measuring instruments.

In contrast, I remember driving a standard family sedan down a tight and windy road that descends a mountain. There were four hefty people in the car, and I was driving fairly hard. The brakes could be easily felt to be fading – pedal pressure went up and up to gain the same retardation. This would have been an excellent test to repeat with upgraded brakes.

In short, I think for a road car, the brakes are best tested using the criteria of 'fade' and 'feel.' But before even thinking of upgrading brakes, ensure that the suspension and tyres are in good condition.

Chapter 6
Testing aerodynamics

Think 'aerodynamic testing' and most people immediately conjure-up visions of expensive wind tunnels. But in fact, highly effective aerodynamic testing can be carried out on the road. One of the most important tests is flow visualisation – literally, being able to see the pattern of airflow over your car. Let's start there.

SEEING FLOWS

As a car moves forward, air passes over its body. But, unlike a boat that pushes water aside, in the case of cars, the fluid (air) is invisible. So in most cases people don't really have any idea of the paths that the air takes over, under and around their car – they just guess. But help is at hand: it is a near-zero cost process to wool tuft a car and then drive it down the road, inspecting the pattern of airflow from another moving car or the side of the road.

The idea is deceptively simple: cut up lots of short lengths of wool, selecting a colour that contrasts well with the car's paint colour. Using good quality masking tape (good quality so that it won't lift the paint when you take it off!), stick the tufts all over the car, keeping them far enough apart that they can't touch each other. (If they're too close, they tend to stick to one another.)

Once you've done that, drive the car (or have someone else drive the car) down an empty, multi-lane road at about 70-80km/h (~40-50mph). From another car, shoot video or still pics of the tufts, including close-up details and overall shots. If you are concerned that the buffeting of the chase car will upset the target's aerodynamics, get further away and use a telephoto lens. You can also photograph from the road-side, although to do this you'll usually need a good camera.

TYPES OF AIRFLOW

Airflow over cars can be basically split into two types: attached and separated. In attached flow, the air is guided by the shape of the car's bodywork. Where this type of flow is present, the wool tufts will line up end to end, the ends fluttering just a little. On the other hand, where the flow is separated, the movement of the tufts is nearly random – they'll wave around in the air, sometimes writhe upwards at 90 degrees to the surface that they're stuck to, and quite often even angle *forward* into the direction of flow!

At the front of a modern car, the flow across the surface of the bonnet will be almost always characterised by attached flow: the tufts will be lined up beautifully. The transition up the windscreen will also be attached, but the transition from the windscreen to the roof may not be so good. If the flow detaches itself at this transitional change of angle, there will be turbulence across the leading edge of the roof. The tufts here will be whirling around, not lying flat and nearly still.

SPEEDPRO SERIES

The two different types of airflow can be seen in this wool tuft testing picture. Across the bonnet (hood), up the windscreen, and onto the roof can be seen attached flow. Directly behind the front wheel on the side of the car is separated flow.

On a sedan, the airflow will generally be attached at the trailing edge of the roof – but then it has to make the transition onto the rear glass. It's the back of the car which is most important in determining overall lift and drag, so it's really important what happens here. If the flow remains attached right down the back window and onto the boot (trunk), the car's doing well. This is because the longer the flow stays attached to the body, the lower the drag. Early separation (that is, before the trailing edge of the car) increases drag.

But what about lift? Here the story is a bit different. If you still want to be able to fit people in the cabin, the centre part of the car

Here flow separation can be seen across the front window glass. Note that one tuft was even pointing forwards at the instant the photo was taken.

TESTING AERODYNAMICS

will be much higher than the rear proportions – so the airflow will have to wrap up and over curves. An aircraft wing generates lift by having a curved upper surface and a much flatter, lower surface – and in much the same way, a car body generates lift as well.

So, the more attached the flow is from the front of the car to the rear, the lower the drag. But shapes like this invariably generate lots of lift because of the flow wrapping itself around those upper curved surfaces. Even worse, if the curved surface is at the very rear of the car (eg the classic Porsche 911 shape), that airflow will generate both lift *and* drag. Instead, you want the airflow to depart the car body without wrapping over any final curves – the reason for raised lips and sharp changes of angle on the trailing edge of the boots of modern aero sedans, and the roof extension spoilers of hatches. However, when the air finally does separate from the body, an area of disturbed air will be created behind the car – the wake.

That's a lot jammed into a few paragraphs, but where does it leave us?

By looking at the patterns of airflow revealed by wool tufting, you can accurately estimate:
- the size of the wake – usually, smaller equals better;
- where the attached flow separates from the body.

Using some rules of thumb, you can then estimate:
- where major lift is occurring;
- where spoilers and wings should be placed to be most effective;
- where minor mods might decrease turbulence.

By placing the tufts appropriately, you can also clearly see airflow patterns through:
- intake ducts (eg oil cooler intakes)
- outlet ducts (eg ventilation exits).

The NHW10 Prius had a drag coefficient of 0.29 – excellent for the time. However, could flow over the rear window be improved? (Courtesy Toyota)

Further, you can generalise with regard to:
- the areas of low pressures (attached airflow wrapping around upper curves and, to a lesser degree, within wakes);
- areas of high pressure (eg the area at the very front of the car above which the flow goes over the bonnet and below which it goes under the car – this is called the stagnation zone).

WOOL TUFTING TO TEST MODIFICATIONS

I carried out testing on an NHW10 Toyota Prius – a sedan, unlike later Prius models. I wondered what the airflow pattern was like on the rear window and boot (trunk) – a common area for flow separation to occur. The first step was to track the airflow pattern over the rear window, using wool tufts. This testing showed that there was attached flow across the transition from roof to rear window (good!). The attached flow continued down the window at both ends of the rear glass; however, in the lower-middle area there was separated flow (bad!). In other words, a separation bubble formed at the middle/base of the rear window, which would adversely affect the flow onto the bootlid.

One of the AirTab vortex generators used to alter the airflow pattern on the Prius. Vortex generators can reduce flow separation.

SPEEDPRO SERIES

Tuft testing of the standard Prius showed that there was attached flow from the roof to rear window. The attached flow continued down the window at both ends of the rear glass, however a separation bubble formed at the middle/base of the rear window (ringed).

The car with four AirTab vortex generators positioned across the roof just ahead of the rear window. (They are hard to see because of the glare on the roof). With these in place, the airflow down the middle of the rear window now stays attached. However, either side of the AirTabs' path of influence, the separation remains.

Six vortex generators are now in place. The airflow pattern is transformed, with no separation bubble forming at all – just some separation at each side of the rear window.

To see if the separation bubble at the base of the rear glass could be eradicated, four AirTab vortex generators were centred at the trailing edge of the roof. With these in place, the difference in airflow was immediately apparent. This time, the airflow down the middle of the rear window remained attached to the glass. This change in flow pattern was directly downstream of the vortex generators. However, either side of this path of influence, the separation remained.

Another two vortex generators were then added, giving a total of six centred on the trailing edge of the roof. Again, the difference was obvious. The airflow pattern was completely transformed, with no separation bubble forming at all. However, with such good airflow, any separation became more visible and some could be seen at the base of the window at each extreme end. Would fitting another two vortex generators (so extending the line across the whole width of the roof) fix this problem?

The answer was 'no.' With eight vortex generators placed on the roof, the separation at the lower edges of the rear glass remained. I decided to go back to six vortex generators and run with the small edge separation remaining.

TESTING AERODYNAMICS

MEASURING AERODYNAMIC PRESSURES

As with the flow testing of intake systems (covered in Chapter 4), the small pressure variations occurring on a car's body can be measured with either a water manometer (near zero cost to make) or a Dwyer Magnehelic gauge. Magnehelic gauges are designed to measure both positive and negative pressures, and so have two measuring ports. By using both ports simultaneously it's easy to measure pressure differentials – just what is wanted in many applications.

When buying a Magnehelic gauge for this application, select a gauge that measures up to a maximum of about 3 inches of water. (The 3-inch gauge lets you use it in other ways as well. If you intend using it purely for aerodynamic work, buy a 0 to 1-inch gauge.) In this application I recommend the Magnehelic gauge rather than the water manometer. We'll start by looking for areas of high pressure – perfect for the intake to the engine.

Firstly, buy 4-5m of small diameter plastic hose that fits tightly over the Magnehelic gauge's high pressure hose nipple. Then, making sure that the mouth of the hose is placed at right-angles to the direction of airflow, place the open end of the hose at the location that you are investigating. Run the other end of the tube back into the cabin, holding the tube in place with pieces of masking tape. All that you then need to do is to drive the car at constant speed, and have an assistant read off the gauge. Once you have measured the pressure at one location, move the hose and repeat the process, making sure that you are doing the same speed each time the measurement is made.

I used this technique to optimise the location for the engine intake on

A 0-1 inch of water Magnehelic gauge is good for measuring small aerodynamic pressures.

an Audi S4 I once owned. On the Audi, a large front-facing duct fed an oil cooler. It seemed reasonable to suppose that in front of the oil cooler a high pressure would be developed, and so if a new intake duct was fed from this location to the engine's airbox, some ram air effect would occur.

But was the air pressure in front of the oil cooler actually high? Testing with a Magnehelic gauge showed that at 80km/h (50mph) there was a positive pressure of 2 inches of water, and at 100km/h (62mph) this had risen to 3 inches of water. So yes, it was a good place from which to pick up intake air. You'll remember from Chapter 4 that the full-load pressure drop across an air filter is typically in the range of 1-2 inches of water. Therefore, in the case of the Audi, by picking up air from the high-pressure location, I could more than make up for the pressure drop through the air filter!

With some further modifications to the intake system to improve its flow, I actually decreased total pressure drop from 32 inches of water to just 9.6 inches of water. The pressure drop of the intake snorkel was reduced from a maximum of 8.4 inches of water to zero – the positive pressure being developed in the enlarged duct exactly making up for the pressure drop caused through friction. And even better than that, at anything less than full throttle, a measurable positive pressure could be seen in the airbox. And how much positive pressure? Up to 9 inches of water at high cruising speeds like 140km/h (87mph)!

BONNET (HOOD) VENTS

Pressure testing can also be used to identify the best location for bonnet (hood) scoops and vents.

For an effective bonnet (hood) vent (ie one where air travels out from the engine bay to the atmosphere) the pressure under the bonnet needs to be higher than the pressure above the bonnet, *at the point at which the vent is mounted.* That is, the best site for the vent is where there is the greatest pressure difference.

I carried out some testing on a Nissan Maxima turbo. Measurements were first made at various locations on the surface of the bonnet (low pressures) and then at various locations on its underside (high pressures). (Note that when measuring low pressures, you need to swap the hose to the negative pressure measuring port on the gauge.) The following table shows the pressures above and below the bonnet, and then the difference. As the table shows, the greatest difference between the underbonnet and overbonnet pressures was at the very leading edge of the bonnet, where it was 0.6 inches of water. However, in fact it was very hard to site a vent here – moving backwards a little to the front third of the bonnet still gave 0.4 inches of water pressure difference at the chosen vent site. The installed vent worked very well.

SPEEDPRO SERIES

	Leading edge of bonnet	Front third of bonnet	Midpoint of bonnet	Rear of bonnet
Above bonnet pressure	-0.5	-0.3	-0.1	0.6
Below bonnet pressure	0.1	0.1	0.3	0.4
Difference	0.6	0.4	0.2	-0.2

PRESSURES AND FLOWS

When we're talking about airflow, we need to keep in mind that it is a variation in pressure that causes airflow. So flow occurs through an air/air heat exchanger only because there is a higher pressure on one side than the other. And, if the pressures each side are the same, no airflow movement through the heat exchanger will occur. The good news is that we can directly measure those pressures, and so see exactly what is going on.

SCOOPS

Direct pressure measurement also works well when evaluating scoops – for example a scoop that feeds air to a top-mount intercooler. I carried out testing on a Peugeot 405 SRDT turbo diesel, a car that instead of using a bonnet (hood) scoop has a duct integrated into the underbonnet sound insulator to feed the top-mount intercooler.

Exactly where on the Peugeot the duct picks up air from isn't all that clear, but it appears to gather air from the gap between the bonnet (hood) and its locking platform. However, as indicated above, it's the pressure *difference* that matters. So even if the feed duct was pretty poor, perhaps the engine bay was optimised to create a low pressure below the intercooler?

The pressure probes were in turn placed in the middle of the top and bottom faces of the intercooler. At a test speed of 100km/h (62mph), there was a pressure on the top surface of the intercooler of positive 0.4-0.5 inches of water (the 0.1 fluctuation being caused by wind gusts and the presence of other vehicles). However, under the intercooler, at the same speed and in the same conditions, there was a pressure of 0.4 inches of water. (The under-bonnet pressure fluctuates less as wind gusts and the presence of other vehicles have less impact.) That meant the difference in pressure above and below the intercooler was just zero to 0.1 inches of water!

It doesn't sound like much of a pressure difference – and it isn't.

As a comparison, the pressure difference across the Peugeot's radiator/air con condenser at the same speed was a relatively constant 0.25 inches of water – 2.5 times as much.

To put this another way: at speed, the airflow through the underbonnet intercooler was terrible. Therefore, installing a larger intercooler core would have achieved little. In this car, the first step in improving intercooling efficiency would need to be the creation of a greater pressure on top of the core (eg by an external bonnet scoop), or the reduction in pressure under the core (eg by experimenting with different shaped undertrays). This measurement (comparison of the pressures top and bottom of the intercooler) should be done on all cars featuring engine bay intercoolers fed by bonnet (hood) scoops.

Using a different undertray to enhance intercooler flow was in fact done on another of my cars. The car was the Nissan Maxima referred to above, a vehicle that

The Maxima front undertray. It improved airflow through the intercooler mounted in the engine bay, and fed air via a scoop.

TESTING AERODYNAMICS

had an underbonnet intercooler installed. It was fed air by a large scoop. The desire to make some aerodynamic changes to the front of the car came about because when doing some other road testing, I'd had the standard undertray off the car. With the undertray removed, the measured intake air temperature rose, indicating that the intercooler was performing more poorly. It therefore seemed that some tweaking of the undertray had the potential to dramatically improve intercooler efficiency.

Pressure measurement showed that, despite the bonnet scoop, the pressure under the intercooler was higher than the pressure on top – that is, the airflow was going from the engine bay, through the intercooler and then out of the 'scoop' (that was really a vent)! In fact, the pressure differential across the intercooler was *minus* 0.1 inches of water.

However, by fitting a custom-made undertray, the pressure differential across the intercooler core was increased from -0.1 inch of water to +0.3 inches. Measured post-intercooler, intake air temps in cruise on a 30°C day (86°F) dropped from 65°C to 47°C (149°F to 117°F).

TESTING LIFT AND DOWNFORCE

Lift and downforce can be measured accurately on the road, and without it costing a lot. You will need to do some careful testing (ie keeping constant the atmospheric conditions, speed and car loads), install some specific sensors, and do a little bit of simple electronics work – but it is straightforward to do. The approach is based on the idea that if there is aerodynamic lift occurring, the ride height will increase. If there is downforce occurring, the ride height will decrease.

Suspension height sensors can be used to accurately measure changes in ride height caused by aerodynamic lift or downforce. These are the pot-based sensors from a Range Rover that I use. Note the white dots I have put on the sensor bodies – the sensors are linear only within this angular range.

The amount that the ride height will increase or decrease is dependent on two factors: the stiffness of the suspension and the amount of the lift or downforce. The softer your suspension, the greater the change in ride height will be at speed for a given aerodynamic force. If your suspension is soft, it will be easier to measure lift or downforce. Harder suspension will extend or compress by a smaller amount for the same aero forces.

However, the difficulty is in this: how do you separate normal suspension movements over bumps from the aerodynamic change in suspension height? And before we even get to that, how do we even measure ride height? The good news is that ride height sensors are readily available – they're used in cars with air suspension. The sensors are connected between the car's bodywork and the suspension arm or axle, and thus constantly measure the height of the suspension.

The best used suspension height sensors that I have found are the Dunlop units that were fitted to P38 1994-2001 Range Rovers. These sensors comprise potentiometers (pots), where the arm connected to the suspension moves, causing the pot to rotate. Unlike more complex designs, these sensors use a simple three-wire design, allowing them to be fed a regulated voltage and then output a varying voltage signal. If buying these, ensure you pick the ones that come with a short section of loom and a specific plug. You can then cut off the plug and hard-wire the sensor into place, or add a new plug and socket. (More on wiring in a moment.)

As standard, these Range Rover sensors have a rod-shaped arm with a 90-degree bend at one end. I cut off the bend and then use a 6mm die to thread the rod. (A ¼-inch die can also be used.) Threading the rod allows miniature ball-joints to be screwed onto the arms. Brackets and links can then be made that allow the sensors to connect to the suspension arms. There must be no 'slop' in this linkage; you need to use quality parts and do a good job if you are to get accurate readings.

These sensors need to be installed so that their movement stays within the linear region – about 110° of rotation. The amount of movement that the sensor arms make through the full suspension travel depends on the leverage ratios in the links. This is best determined through trial and error. Ensure that the ball-joints do not bind at any suspension position – remove the springs, and use a jack to move the suspension when testing for this.

You can fit two sensors, three sensors or four sensors. The minimum should be one sensor at the back and one sensor at the front. If you mount the sensor so that it is rotated by the central part of an anti-roll bar, an average of the left and right suspension movement at that end of the car will be gained. However, in some cases it is simpler to mount the sensor on just one wheel at the front and one wheel at the back – normally, the aerodynamic lift or downforce is symmetrical from side to side of the car.

Do not underestimate the task of fitting such sensors to the front and rear suspension. It's not difficult, but it is fiddly and tends to take much longer than you expect. I don't recommend using a conventional pot and rigging up something temporarily – water and/or dust will wreck the open pot quickly and if you are making aerodynamic modifications aimed at altering lift, you will want to be able to undertake testing over a period as you refine the modifications. The resistance track on a normal pot will also wear out quickly.

With the pot fitted, we now have a sensor that outputs a voltage which varies with ride height. So when the car is stationary, the voltage output of the sensor on the rear suspension might be 2.3V. But as soon as we start driving down the road, the voltage coming from the sensor is going to dance all over the place as the suspension springs deflect with bumps. What we need is a way of showing the average voltage output of the sensor, say with the voltage averaged over 5 seconds. This sounds complex to achieve, but is actually quite easy – and cheap too.

You will need a multimeter (to measure the voltage coming from the averaging circuit), a 5V regulated power supply that will operate off the car's 12V system (to feed the circuit, including the sensor), a 220µF 16V electrolytic capacitor, and a 10 kilo-ohm potentiometer that is wired as a variable resistor.

The circuit is shown in Figure 6-1. Note that I have not shown the 5V power supply in detail – simply buy an eBay module capable of outputting 5V (only a low current is needed) and operating off the car's 12V supply. Note also that the capacitor is polarised, with its negative terminal marked by a line of negative symbols down the body, and so it must be connected into the circuit the right way around.

In use, rotate the circuit's pot (variable resistor) until the averaging period is as you want it. This can be tested by pushing the car up and down and then seeing how fast the multimeter display returns to a static reading when you stop pushing. I think about 5 seconds works well.

On the Range Rover sensors, the wiring colour codes are:
- Brown – Ground
- Blue – 5V
- Green – signal

With a 5V feed, rotation over the linear range gives a 0.8-4.5V output.

A Range Rover suspension height sensor being installed on the front suspension. Note the use of the two miniature ball-joints and the additional link. Installation is fiddly but once installed, the sensor can be used to provide you with a continuous readout of ride-height. There must be no slop in the linkage!

TESTING AERODYNAMICS

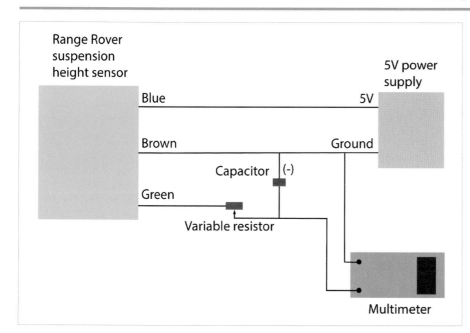

Figure 6-1: This simple circuit can be assembled very cheaply using eBay parts. It averages the output of a ride height sensor, allowing bumps and road imperfections to be ignored but still allowing you see even small changes in ride height caused by aerodynamic lift or downforce. The variable resistor can be set to give the desired averaging time. A multimeter is used to monitor the sensor output.

Ensure that the voltage increases with increasing ride height, or you could be measuring lift and think that you are measuring downforce! If the voltage falls when you want it to rise, swap the brown and blue wires.

The system can be used in a number of different ways. Simplest is to leave the output in just the raw voltage. The change in voltage is indicative, then, of the change in ride height, reflecting lift or downforce. For example, if the voltage measurement of the front suspension is 3.70V at normal ride height, and 3.55V at 100km/h (62mph), the suspension has decreased in height by '0.15V.'

With the car stationary again, you can add weights to the car at that axle line until this change in voltage is replicated – this will show you how much downforce was being developed. (Or you can use a spring balance – or digital crane scale – to provide the equivalent lift.)

I have fitted my Honda Insight with custom air suspension. The control system uses three of the previously described Range Rover

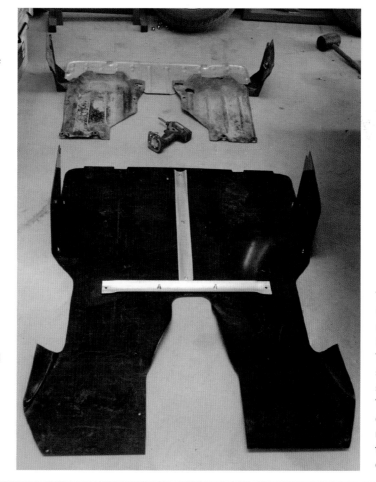

The front undertray I made for my Honda Insight. The original standard one is shown behind. This undertray developed measurable front downforce.

SPEEDPRO SERIES

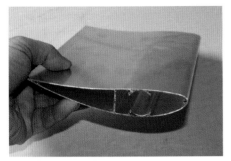

By measuring average suspension height at speed, you can see if a wing is actually providing downforce. Furthermore, you can also best tune its angle of attack.

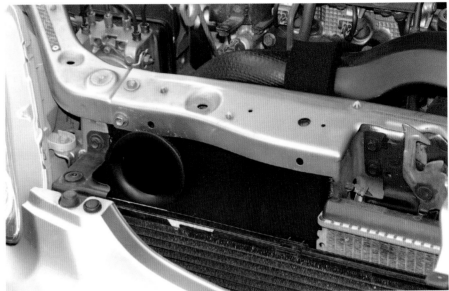

A new engine intake, positioned in a location of measured high aerodynamic pressure.

height sensors, feeding signals to two dedicated suspension controllers. These vary suspension height via air solenoids. I also have 'live' readouts of front-right, front-left and combined-rear suspension heights (reading in millimetres height above the bump stops) on my MoTeC dash. I use a running averaging time of 5 seconds for these displays.

With the Insight aerodynamically standard but for a large, extended undertray at the front, the following results were gained. The rear ride height did not change with speed. On the other hand, the front ride height clearly decreased, indicative of front downforce being developed. At 110 km/h (68mph), I consistently saw a decrease in front ride height of 5mm (just over 3/16 in). To statically achieve this suspension deflection required the placement of 15kg (33lb) on the front axle line, so at this speed, 15kg (33lb) of downforce was being developed.

Aerodynamic testing is one of the most important areas in which you can test. The airflow through intercoolers, oil coolers and radiators is dramatically influenced by aerodynamics. The stability of a car on the road is influenced by lift and downforce. The fuel economy (mileage) of a car is substantially impacted by aerodynamic drag. The testing techniques covered in this chapter allow you develop and test your modifications in each of these areas. Once you've started doing so, you'll wonder why you never did it before!

Using wool tufts to show the airflow pattern near the side mirror. This car, an XE Jaguar, has excellent attached flow in this area.

Chapter 7
Programmable engine management

If you have a programmable management system fitted to your car, can you test and tune it on the road? The answer is yes, you can – with some caveats. Firstly, the car cannot be immensely powerful. It's just too dangerous if full throttle results in mass wheel-spin, or tuning at higher rpm in fourth, fifth or sixth gears results in warp speed. Secondly, you need a mix of roads – from heavy urban traffic all the way through to near-empty blacktop.

Tuning programmable engine management is a subject that deserves its own book, so here I'll concentrate only on some of the aspects that tuning on the road creates.

ON-ROAD TEST AND TUNE TOOLS
To tune and test on the road you'll need at least four tools.

1. Wide-band air/fuel ratio meter
A wide-band AFR meter is a must-have. Without one, you're very much risking blowing up your engine through overly lean mixtures, and at a minimum, you'll chase your tail a lot when tuning. Having an accurate indication of air/fuel ratio is critical to effective engine management tuning and testing.

A good quality air/fuel ratio meter is a must-have if you are testing and tuning programmable engine management on the road. (Courtesy Innovate)

It is far better if the sensor is screwed into the exhaust, close to the engine. Yes, you can tune with it up the tail pipe, but response will be slower (important when setting things like acceleration enrichment) and the reading will be a little less accurate if the sensor is located after the catalytic converter. Buy the best wide-band AFR meter that you can afford, and then look after it.

2. Knock detection system
Unless you are using software that works with the factory engine management, and can therefore monitor the factory knock sensor (and work with the factory ECU knock sensor logic), you need a way of detecting engine knock (detonation).

The cheapest, most effective and simplest way of doing this is to use a remote-mounted microphone in the engine bay, connected to an adjustable in-cabin amplifier. The person listening to the

SPEEDPRO SERIES

engine should wear fully enclosed headphones – and that's especially the case if the car has a loud exhaust.

It's easy to make your own listening system, starting with a cheap eBay amplifier and then extending the microphone wiring and adding good-quality headphones – more on how to do this in a moment.

3. Sparkplug spanner
Periodically when tuning, you should pull a sparkplug and inspect it closely. Plenty of charts exist on the web that will show you what sparkplugs look like when the engine is detonating (for example, tiny globules of aluminium deposited on the porcelain insulator), or when the plugs are overheating (erosion of electrodes), or when the mixtures are rich or lean – and so on. Being able to remove and carefully examine sparkplugs is an important element in tuning. 'Reading' sparkplugs is an art – do an online search to find some interesting coverage of the techniques.

4. An assistant
You also need an assistant – either to drive the car or listen to the engine and operate the laptop. Note that the assistant doesn't actually need to know a lot about tuning! My on-road tuning approach is to drive the car, feeling its behaviour and watching the AFR display, while someone else operates the laptop. Based on what I can see, hear and feel, I ask for specific tuning changes. Taking this approach, the assistant obviously needs to be able to find their way through the maps and make the required changes, but a detailed knowledge of tuning isn't needed.

BUILDING AND USING A KNOCK DETECTION SYSTEM

If you are testing and tuning programmable engine management on the road, you must use a detonation (knock) detecting system. Here's how to make one.

The design is based on the 'Listen Up Portable Personal Sound Amplifier' that you can find from lots of eBay sellers at a very low price. You'll also need to add a decent pair of earphones or headphones, four metres of cable (eg shielded two-core microphone cable) and a battery clip. In addition, you'll need to have the skills and equipment to do some simple soldering.

The Listen Up Portable Personal Sound Amplifier – available cheaply on eBay – is the basis of the detonation detection system. It comprises a compact, battery-powered microphone and amplifier.

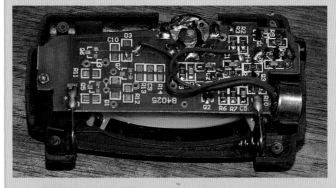

Open the box by first taking off the end caps (one covers the battery – a single AAA cell) to reveal four small screws. With it open, it should look like this.

Note the microphone (arrowed) – both the fact that it is a large and relatively good quality unit, and that it is attached to the printed circuit board with flying leads. Unsolder the microphone and solder in place the new long shielded cable.

PROGRAMMABLE ENGINE MANAGEMENT

The new wires soldered in place. I also soldered the braid of the cable to the negative connection of the battery (arrowed) to provide better shielding of the signal. Make a suitable hole for the cable to escape and then close up the box.

Solder the other end of the microphone cable to the microphone. Keep the polarity the same as original connections.

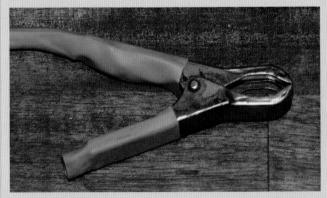

Use hot melt glue or similar to mount the microphone to the inside arm of a metal battery clip; then cover in heatshrink.

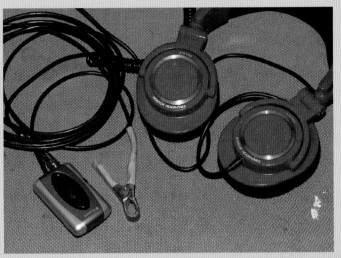

High quality, fully enclosed headphones will work best – but any decent quality earphones should also be fine. With fully enclosed headphones worn, the sound quality is excellent.

Using the detonation listener is very simple. You clip the microphone to whatever you are interested in listening to. Noises are transmitted through the metalwork directly to the clip and microphone, making the device extremely sensitive.

To detect detonation, the clip is best placed directly on the block, in the type of place that the factory knock sensors are positioned – no surprise there! Adjust the volume control to give a comfortable loudness level, then sit back and listen. Over the clatter of pistons, valve gear and gearbox whines, detonation sounds like a sharp 'splat, splat!'

However, the device's functions are not limited to this alone. Setting idle speed is easier when listening to the engine because you can better hear hunting; you can use it to ensure that over-run injector shut-off is working, and you can also use it to detect if on-off vacuum solenoids are operating as they should.

Note that it's best to listen from the passenger seat while someone else drives the car – that way, the driver can still hear emergency vehicles and concentrate on driving, rather than listening to strange noises …

For its cost and effectiveness, this is one of the best on-road tuning and testing tools you can have.

SPEEDPRO SERIES

Listening for detonation on the dyno (rolling road). Don't underestimate the power of human ears in knock detection. With the portable electronic listening system covered on the previous pages, this can easily be done on the road.

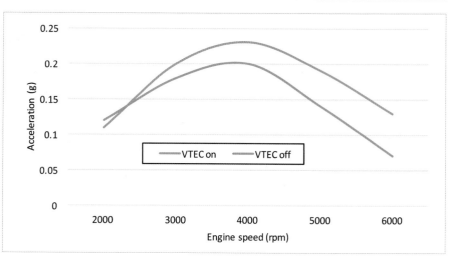

Figure 7-1: Working out where to place the switchover point for VTEC adjustable cam timing. One run was done with VTEC switched off, and the other with it switched on. The revs at which the lines cross show the best full-throttle changeover point.

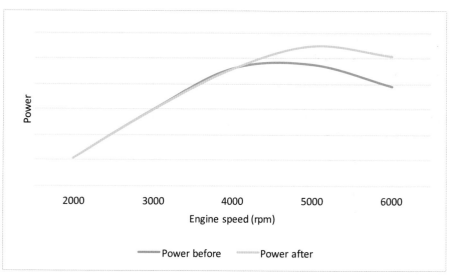

Figure 7-2: The results of revising top-end boost, ignition timing and fuelling on a turbo car with programmable engine management. Being able to measure this information at no cost by on-road testing is very useful.

MEASURING POWER WHEN TUNING

In Chapter 2 I introduced the approach of using an accelerometer to measure changes in engine torque and power. The approach is particularly good when you are tuning programmable management. Let's take a look at a few occasions when I did just that.

First up, I was tuning my little Honda that uses VTEC changeover valve timing. It's the VTEC-E system, so not one aimed at radical increases in power, but still a system that does change breathing capability substantially. So at what engine speed (rpm) is it best to switch the system?

Finding this out is as easy as doing two lots of measurements – the instantaneous accelerations across the full range of engine revs with the system switched off, and then the same with the system switched on. Where the lines cross is the best changeover point. Figure 7-1 shows the results. As can be seen, on this car the best changeover point is at very low rpm – around 2200rpm.

When tuning something like VTEC-E, there's another point to consider. Rather than map the changeover point against just rpm, you can use a 3D map that also maps it against throttle position. To best find this calibration data, you can do the acceleration testing at different, fixed throttle positions. The easiest way of doing this is to temporarily place differing height stops under the throttle (eg blocks of wood). For example, you could then do the testing at one-third, two-thirds, and full throttle. From this family of curves, you can work out what cam timing switching strategy works best at different throttle angles.

PROGRAMMABLE ENGINE MANAGEMENT

You can also take the above approach when tuning a variable intake manifold.

Another example of where I measured engine torque and power was when tuning the top-end air/fuel ratios and ignition timing. When tuning a small, high compression engine, one never designed for turbocharging (and running 15psi boost!), you need to be conservative. But after tuning the car for a while, I decided that I was probably being *too* conservative. Figure 7-2 shows the results, measured on the road, of revised top-end boost, ignition timing and fuelling. There was about a 20 per cent power gain at 6000rpm.

Your main on-road tools when testing and tuning programmable engine management are your laptop's displays, your listening system for detonation, and a good quality air/fuel ratio meter. But don't forget that having the ability to plot power and torque curves allows you to easily quantify gains and losses you might be making. Also, when tuning 'variable' systems like variable intake systems and variable cam timing, being able to measure on-road instantaneous acceleration makes optimisation much easier. Finally, and I haven't tried it yet, I think on-road tuning could be very useful when setting-up an electronic throttle control system.

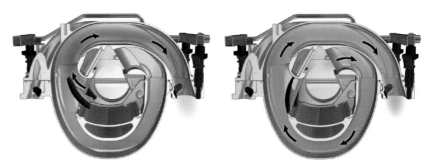

The best rpm at which a variable intake manifold should be switched from long to short runners can be found on the road by plotting the power or torque curves in each configuration. (Courtesy Mercedes)

Aux Table 2											
	TP %	0.0	10.0	20.0	30.0	40.0	50.0	60.0	70.0	80.0	100.0
AT °C	60.0	100	100	100	100	100	100	100	100	100	100
	50.0	100	100	100	100	100	100	100	100	100	100
	45.0	0	0	0	75	100	100	100	100	100	100
	40.0	0	0	0	75	100	100	100	100	100	100
	30.0	0	0	0	50	50	50	50	50	50	50
	20.0	0	0	0	0	0	0	0	0	0	0
	10.0	0	0	0	0	0	0	0	0	0	0

The 3D table in a programmable management system that controls pump speed in a water/air intercooler system. At low throttle openings and low intake air temperature, the pump stays off. However, depending on temperature and throttle, the pump voltage can then rise to 50 per cent, 75 percent or 100 per cent. Mapping a table like this is best done on the road by monitoring intake air temperature. This the fourth iteration of the table, that has been tweaked over a year of on-road driving.

Chapter 8
Performance modification: a personal approach

A long time ago I had a new performance car. I was pretty happy with my purchase, but, as is the way of the world, I wanted it to go faster. I didn't then know a lot about performance modifications, but I knew that lots of other people with the same type of car were getting a big exhaust put on it. I booked it into an exhaust workshop – but I also did something that apparently lots of other people were *not* doing. I measured the car's standing-start acceleration time on the road. The car was equipped with automatic transmission, and so I could quite consistently measure the time it took to get to 100km/h (~60mph).

I took the car in and had the exhaust done – a completely new high-flow system from the turbo to the tail-pipe. The car sounded sportier and felt to me like it went harder. Then, to quantify just how much harder it was going, I ran some more standing-start times. *But the times were unchanged*. The new, expensive exhaust had made absolutely no difference to the car's performance!

It was my first – and very salutary – lesson, that lots of people spend money on performance modifications that make no difference. In addition to feeling very crestfallen about the outcome, it also made me start thinking about car modification workshops – did they really know what they were doing?

These thoughts were reinforced about a year later. In the 12 months since the exhaust had been fitted, I'd become really interested in car modification. I'd read and thought a lot about the subject. I decided I needed to fit a big intercooler to my turbo car, and I had a question. An intercooler is really needed only when a turbo car is on boost; at other times it poses a flow restriction. So if I was going to fit an intercooler, should I also fit a bypass so that when off-boost, the intake flow would be less restricted?

I rang a turbo workshop and asked this question. I then lived in a major city, so this was no backyard workshop, but instead one with a big reputation. But the workshop staff had no idea what I was talking about – they didn't even understand the question.

I then rang a college that taught automotive mechanics and asked the same question of an automotives lecturer. At least he understood the question – but he didn't have an answer. He then put me onto another lecturer, one who worked part-time because, for most of the time, he was running his own engine dyno workshop. This time, the man did know the answer. "You'll get no advantage," he said, "because the volumetric efficiency of the engine is so poor with the throttle at less than fully-open, the restriction of the intercooler is inconsequential."

PERFORMANCE MODIFICATION: A PERSONAL APPROACH

It was a good answer, from a very smart man who I got to know well in later years. Those were the years when I changed careers from being a school teacher to an automotive journalist, writing primarily about modified cars. In my new career, I got to visit many dozens of car modification workshops, including the one that I'd (anonymously) asked that question about the intercooler bypass years before. And, as with many very well-known car modification workshops, I found that their knowledge of basic physics was quite poor. One day they told me that their new intercooler was so good that they could record engine intake air temperatures lower than ambient – and no, the intercooler didn't have any refrigeration ability …

I also kept modifying my own cars. In fact, ever since I'd run the acceleration time on the car equipped with the new exhaust, I'd methodically and rigorously tested every single modification I'd performed. If I made a change and there was no widely-known way of testing it (eg the restriction of an intake system), I investigated and tested until I'd come up with an effective way of finding out precisely how well the modification worked.

And having by this stage published many magazine and web stories with literally hundreds of car modification workshops around the country, I'd also formed the view that I wouldn't trust 95 per cent of them with my own cars. There was simply too much guesswork, too much experimentation on customers' cars, not enough research and often, very little thought. So I did 99 per cent of the modifications myself, including some – like modifying re-gen braking on a hybrid car – that were world-firsts.

And I tested and tested and tested. Every modification that I made I wrote about for a magazine; every modification that I said was successful had to be provable as working. And, if the modification didn't work, I also wrote about that.

Cars and projects came and went – one turbo, two turbos, supercharged, naturally-aspirated, diesels, hybrids. Two cylinders, three cylinders, four cylinders, five cylinders and eight cylinders. (One day, I hope for 12 cylinders and pure electric!) And each time I made a modification, I'd test it. Occasionally I'd get a dyno run done, but after a while, I found the testing results I got on the road were much more useful.

And, over time, I found that I'd developed a performance approach that seems different to most people. This is a long-winded introduction to this, the last chapter, but it also shows where I'm coming from. So I'd like now to look at the performance modification approaches I take.

INTAKE SYSTEMS

Modifying the intake system can result in some appreciable power gains at almost no cost. And, unlike exhausts, the intake changes can be easily made working at home with just hand tools. Power gains will vary substantially – it will depend on how good the original system is and how good the modified system is. However, at the extreme, I have seen power gains of five per cent made by just fitting a better intake duct to the standard airbox.

A number of different approaches can be taken when modifying intakes, but my preference is to retain the standard airbox and simply make sure that lots of air from outside the engine bay can get to it with a minimum

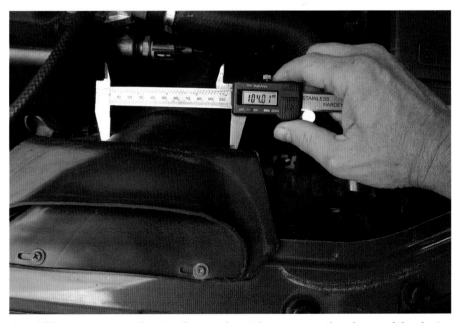

In addition to pressure-drop testing, work out the cross-sectional area of the ducts used in the intake system. Smaller equals worse flows, and bottlenecks are not at all uncommon on standard intakes.

of restriction. Taking this approach has advantages that are easily summarised – it's cheap and it's usually very effective. Retain the standard factory filter element in the box – the huge power gains apparently made by replacing this element with an aftermarket one are simply mythical. (Just measure the pressure drop across the filter to see that for yourself!)

Investigate how the factory intake to the airbox works. Do this in three ways:
- inspection
- intake manifold pressure-drop test
- intake air temperature measurement.

Does the intake system pick up air from inside the engine bay (bad, because most of the time, this air will be hot). Does it have a duct that breathes air from near to a headlight or an opening in the grille (good, because this air is cool). Does the duct pass through a resonant box (bad, it'll cause restriction) or is a resonant box tee'd into it (not so bad). Is the mouth of the duct flared out (ie bell-mouthed) so that it will pick up air more effectively, or is it just a sharp-edged ending, cut off square? Is the mouth of the duct pushed up hard against an obstruction?

What is the total pressure drop of the system? A maximum of 10 or so inches of water is going to be hard to improve on, whereas if it's at 25 or 30 inches, you can be certain that major improvements in flow will be possible. Based on your inspection and measurement, you might decide to retain a portion of the factory airbox intake, or to replace it entirely.

The use of 75mm (3in) PVC plastic stormwater pipes and fittings makes for an effective and easily fabricated intake duct. The end of the pipe can be flared into a bellmouth if it is heated with a heat-gun until the plastic is just pliable, and then the pipe is forced down over the outside of a funnel or over a round-bottom, inverted china bowl. The pipe can be adapted to the airbox intake, with off-the-shelf sections that change the shape from round to rectangular, or the intake hole in the airbox can be enlarged to suit the new duct. Paint the finished item with a can of black spray paint and it will all look fine too.

Don't have the mouth of the intake positioned close to the road. In this position it will pick up dirt and small stones, and the stones can be breathed-in with such speed that they'll penetrate a paper panel airfilter. Also, water will be easily ingested with a low air pick-up; it's not at all unusual to hear of people who have hydraulic'd their engine through this technique. Rain isn't usually a problem – most airboxes have a drain hole in their base and it's easy to put some others in along the length of the duct. You can't make an intake duct to an airbox too big, so if you're looking at a powerful engine, the use of more than one intake duct may be necessary. Note that a free-flowing duct will normally pick up more dirt than standard, resulting in the need for the filter element to be changed more frequently than the manufacturer recommends.

If the airbox in your car is small (resulting in a small panel filter area) or both the inlet and outlet ducts are crudely designed, you may wish to replace the complete airbox with a better one from another car. If you find obvious flow restrictions within the standard airbox, but space or other considerations mean that you can't swap it for another, you can modify the standard box. Smoothing sharp exit and entrance radii, the use of a plastic bellmouth in the exit duct, and removing plastic ridges and strips located in the wrong places will all help flow. Smoothing is easily done with a file and then fine emery paper.

Here I have used the airbox from a completely different car. Note the intake to the airbox that draws air from in front of the radiator, an aerodynamically high-pressure area.

PERFORMANCE MODIFICATION: A PERSONAL APPROACH

A new airbox, pictured during fabrication. It used a large cylindrical airfilter, normally fitted to a Saab 9000. Show here is the intake to the airbox, where I used an ex-subwoofer flared port as the intake bellmouth.

While thankfully it's now much rarer than it used to be, there are still people who remove the whole airbox and intake pipe and place a filter directly on the throttle body. (Obviously, most often on engines that use MAP sensing not airflow meters.) This has two major disadvantages – hot air is breathed all of the time, and secondly, the resonant tuning that the manufacturer has made of the duct between the airbox and throttle is lost. Testing I have performed has shown that playing around with this part of the system can massively vary lower rpm torque – try to keep the volumes and length between the intake box and plenum chamber similar to standard. (Having said that, if you're after every little bit of power, modifying intake lengths and volumes and then running on-road accelerometer testing is a fertile area for low-cost improvements.)

EXHAUST SYSTEMS

One of the most cost-effective mods on almost any car is an upgraded exhaust. While it depends on the car and what else has been done to it, a good rule of thumb on a new high-flow exhaust is to expect about a 10 per cent gain in peak power – enough to make a noticeable difference to real-life on-road performance.

Exhausts are made from a number of materials – stainless steel, mild steel, or aluminised mild steel. Irrespective of the material, all pipes flow the same (the differences sometimes quoted in regard to internal surface finishes can be ignored if the pipe is made big enough in the first place). In actual use, the pipe materials have little to differentiate them, except for durability. If you expect to keep your car for a long time, consider stainless. Otherwise, pocket the substantial savings and go for mild or aluminised steel.

Exhaust pipes can be bent using either mandrel or press bending. Press bending machines are common – every exhaust shop has one. Mandrel bending machines are much rarer, usually being found only in manufacturing exhaust shops. Press bends are slightly flattened, while mandrel bends maintain a similar diameter right around the bend. Making the issue more complex, most exhaust shops and home builders (me included!) that build a 'mandrel exhaust' just get a box of mandrel bends and weld them together. This means that there are 'steps' and often penetrated welding beads at each join – not good for flow. In fact, given the cost difference between the two types of exhaust, it's often best to get a press-bent exhaust in a pipe *one size larger* than is required. That way, it's cheap and flows very well.

In all the factory exhaust systems I have flow tested, the catalytic converter has been the single most restrictive part of a standard exhaust system. This means that replacing the exhaust but leaving the factory cat in place isn't the best approach. In fact, it

I used this ex-Jaguar straight-through resonator as the rear muffler on one of my turbo cars. Working in conjunction with another straight-through muffler, noise levels and flow restriction were both low.

SPEEDPRO SERIES

makes sense to actually have a cat installed that's larger than the nominal pipe size. For example, a 3in cat on a 2½in exhaust will give better results than a 2½in cat on a 2½in exhaust. Testing has even shown that expanding a 2½in aftermarket cat's nipples to 3in results in a good improvement in flow. However, shallow angled cones should then be used to adapt one pipe size to the other, reducing the restriction across the transitions in pipe size.

The best mufflers to fit are large body, straight-through designs. Simply put, straight-through mufflers have nearly zero flow restriction, and good straight-throughs have the best noise suppression as well. Note also that the larger the body of the muffler, the more likely that a given design is to be quiet.

The fitting of resonators (small additional straight-through mufflers) is worthwhile to keep noise down. If you are unsure as to whether a resonator will be needed, ask the exhaust shop to incorporate a straight section of pipe within the exhaust system, so that it's cheap and easy to add the resonator later, should one prove to be needed. I have also achieved a reduction in resonance by simply inserting a larger diameter pipe for a short distance eg 500mm (about 20in) within the pipework.

If you are on a tight budget and have a small car, make friends with the local high-performance exhaust shop. Once you've done that, you'll be rather surprised by how many near-new exhaust systems are thrown away. Some of these exhausts – when bought new from the maker – are worth thousands and include (relatively) high-flow mufflers and cat converters. The components from these systems are ideal for smaller cars, and the cost of the materials can be near-zero.

In most (but not all) naturally-aspirated cars, extractors (headers) are a worthwhile upgrade over cast iron manifolds. However, you need to be a little careful, in that if the cast iron manifold looks good with long runners and (say) a 4-2-1 design, it probably *is* good. Tight, short pipes from each cylinder are likely to flow as poorly as they look. In a naturally-aspirated car with a poor manifold design, about half of the power gain of a full exhaust system comes from the extractors.

In all but one test I have ever performed, the bigger the pipe (and so the lower the back-pressure), the better the power. On the exception I'm willing to bet that if the management system had been slightly retuned (eg low rpm ignition timing advanced) the power at low revs would have picked up over standard as well (it was already improved over standard at high revs). Thus, *after the tuned length section of the pipe is finished* (ie after the first muffler/resonator/cat converter), in a naturally-aspirated engine you should go for a large pipe. In turbo cars, go for a large pipe straight off the turbo.

ENGINE MANAGEMENT

Engine management has changed dramatically over the last 25 years, and so the best approach very much depends on the age of the car you are modifying.

While I enjoy tuning programmable management on the road, where the standard engine management software has been 'cracked,' best results will come from professional remapping of the factory-fitted system. Here my diesel Skoda Roomster is being remapped.

PERFORMANCE MODIFICATION: A PERSONAL APPROACH

Modern cars have engine management systems that are inextricably tied to other car systems (eg the dash, transmission control, etc). If there is software that allows the standard system to be remapped (eg by tuning real-time) then this is the best approach to take. Remapping the standard engine management is also the best approach if the car has to pass emissions inspections, have standard OBD access, etc. If these aspects do not matter, a programmable engine management unit can be used, either working standalone or as a piggy-back.

Over the years I've taken every engine management approach from doing nothing (the air/fuel ratios and ignition timing were fine, even with modifications), to making my own interceptor-style electronic changes, to having the factory engine management system remapped by experts, to fitting and tuning fully programmable engine management (with all the tuning carried out by me on the road). If your car has had its engine management software 'cracked', and if you have access to an expert workshop you trust (one of the five per cent I mentioned earlier!), then I think remapping the factory system to suit the modifications you've made is the best approach. However, if you love doing things yourself, and you want a hobby you can pursue for literally the rest of your life, fitting programmable management and then road tuning it is huge and engrossing fun.

BRAKES

Upgrading brakes is an area where it is easy to get advice that later proves to be (expensively) wrong. To put that another way, people who say: "Oh yeah, just put on 'X' calipers and 'Y' discs" are potentially ignoring a whole bunch of factors that might result in the upgrade proving to be a disaster. Like what factors, then? Try aspects like wheel clearance problems, the need for custom brake caliper adaptors, altered front/rear brake balance, master cylinder incompatibility, overall cost-effectiveness – the list goes on and on.

In increasing level of complexity, cost and risk, here are some potential brake upgrade routes:

CAMS? COMPRESSION? HEADWORK? TURBO SWAPS?

I've also played a bit with internal engine modifications like cams, increased compression and headwork. However, although I think many will disagree, the rate at which you start to spend money, versus the improved car on the road, soon starts to look a bit sick. I can't see myself going back to engine mods of this magnitude – easier to swap in a new and more powerful engine, or add a turbo to an engine that never had one.

A big-brake front upgrade I performed on one of my cars, with the old hardware in the foreground. The new calipers and their mounting brackets are from another car made by the same manufacturer – they were a direct bolt-up. The pads and grooved disc are aftermarket. This was a very effective upgrade that worked fine with the standard ABS and stability control.

SPEEDPRO SERIES

1. Better quality pads, new standard sized discs and new fluid. Rest of system checked and working effectively. Known outcome, low risk and relatively low cost. No legal issues. Best for street driven cars with 20-30 per cent power upgrade.
2. Straightforward factory parts swap, eg braking system components from higher performance version of your car. No change usually needed to brake bias valve or master cylinder. Swap works with standard ABS. May need larger diameter wheels. Known outcome, low risk and medium cost. Few legal issues. Best for street driven cars with 30-50 per cent power upgrade.
3. Non-standard factory parts swap, eg mixing and matching brake parts of different models from a single manufacturer (or manufacturing group). May require altered master cylinder, ABS (sensors, hydraulic control unit and electronic control unit) and brake bias – however, these parts should be readily available. May need larger diameter wheels. Less certain outcome, medium risk and medium/high cost. May have some legal issues. Best for cars where No 2 above not available and with 30-50 per cent power upgrade.
4. Major changes in hardware – aftermarket calipers and discs, component swaps between manufacturers. May require every braking component to be altered, and may need custom fabrication (eg brake hoses, caliper mounts, master cylinder mounts). Will almost certainly require tweaking to achieve desired outcome (eg manually adjustable brake bias), high risk and high cost. Has major legal implications. Best for heavily modified, very high-performance cars.

SUSPENSION

Upgrading suspension can be immensely complex – or very simple. Most people are after increased grip (as opposed to handling, something that includes more than just grip) – and the easiest way to gain more grip is to fit stickier tyres. Going lower profile and wider at the same time will improve steering response, while maintaining the same rolling diameter and so gearing.

The first level of improvement to handling is to change the steady-state *balance*. That is, if the car tends to understeer, to reduce understeer. And if the car tends to oversteer, to reduce oversteer. The easiest way of achieving this is to stiffen in roll the *opposite end* to the one that slides first. So, an understeering car can be improved by stiffening the rear, eg by a stiffer rear anti-roll (sway) bar and/or by stiffer rear springs. If the car oversteers, stiffen the front in roll. (This is a deliberate over-simplification, but some rules-of-thumb make thinking about the topic much easier.)

If stiffer springs are used, the dampers normally must be upgraded to suit. So for example, you may in an understeering car decide to stiffen all four corners, but to stiffen the rear to a higher degree. All four dampers then also need to be stiffer, especially in rebound damping. Dampers strongly influence non-steady-state cornering, eg turn-in, roll linearity, etc.

Over the years, I've done every suspension upgrade I can think of – from just a new anti-roll bar, to all-new springs and dampers and anti-roll bars, to revised suspension arm geometry to – in the case of human-powered vehicles – developing the whole suspension system from scratch. On the basis of that experience, I'd suggest that it's best to start off by making cheap and simple changes like anti-roll bars. (And it's even better if they're adjustable.)

Don't forget that measuring the car's suspension natural frequencies (front in bounce, rear in bounce, roll and pitch) will let keep track of what you're starting with – and what you end up with. I'd also reiterate what I said in Chapter 5, and that is that a skidpan – either formal or a roundabout – is an excellent way of feeling and developing steady-state handling.

Veloce SpeedPro books –

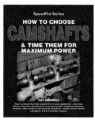 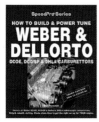     

978-1-903706-59-6   978-1-903706-75-6   978-1-903706-76-3   978-1-903706-99-2   978-1-845840-21-1   978-1-787111-68-4   978-1-787110-01-4

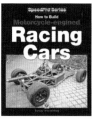   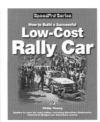 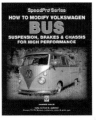 

978-1-787111-69-1   978-1-787111-73-8   978-1-845841-87-4   978-1-845842-07-9   978-1-845842-08-6   978-1-845842-62-8   978-1-901295-26-9

978-1-845842-89-5   978-1-845842-97-0   978-1-845843-15-1   978-1-845843-55-7   978-1-845844-33-2   978-1-845844-38-7   978-1-787113-34-3

 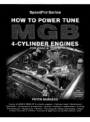     

978-1-845844-83-7   978-1-787113-41-1   978-1-845848-33-0   978-1-787111-76-9   978-1-845848-69-9   978-1-845849-60-3   978-1-787110-91-5

 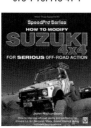 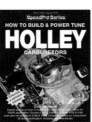    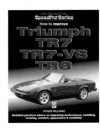

978-1-845840-19-8   978-1-787110-92-2   978-1-787110-47-2   978-1-903706-94-7   978-1-787110-87-8   978-1-787111-79-0   978-1-787110-88-5

978-1-903706-78-7   978-1-787113-18-3   978-1-787112-83-4

– more on the way!

MORE FROM VELOCE ...

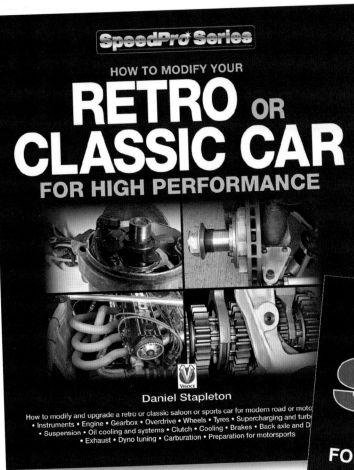

There are four crucial aspects of a classic car's performance: straight-line top speed, cornering speed, acceleration, and braking. This book's detailed guidance shows you how to improve each of these, whether for more enjoyable day-to-day use, or for a diverse range of classic motorsport.

ISBN: 978-1-845842-89-5
Paperback • 25x20.7cm
• 144 pages • 309 colour and b&w pictures

Reprinted after a long absence! This book explores the real off-road potential of each Suzuki model and explains what modifications will give the very best performance for fun or sport. Written by an off road expert, and with 175 colour photos this is an invaluable guide to getting the most fun and performance out of YOUR Suzuki 4x4.

ISBN: 978-1-787110-92-2
Paperback • 25x20.7cm
• 128 pages • colour pictures

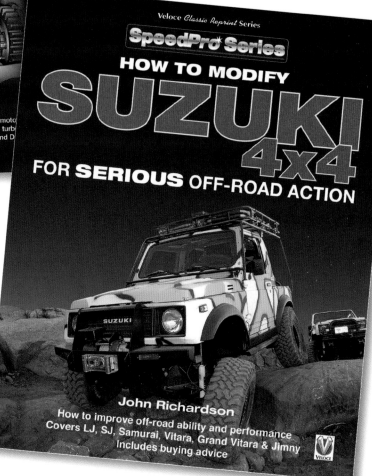

... AND MORE FROM JULIAN EDGAR

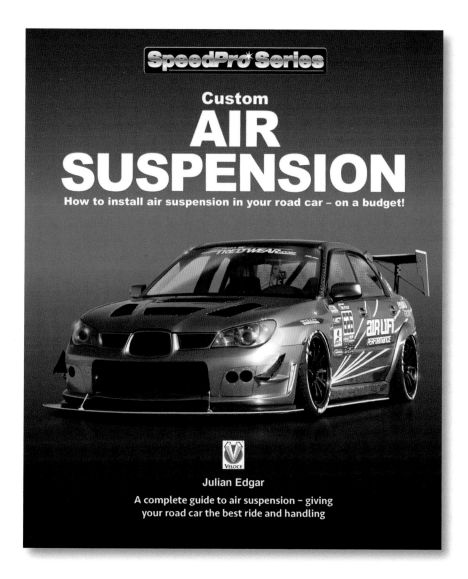

The first book to show you how to fit air suspension to your car. It covers both theory and practice, and includes the step-by-step fitting of aftermarket air suspension systems and building your own with parts from other cars. If you want the best ride and handling for your road car, this is the book you need!

ISBN: 978-1-787111-79-0
Paperback • 25x20.7cm
• 64 pages • 82 colour and b&w pictures

For more information and price details, visit our website at www.veloce.co.uk
email: info@veloce.co.uk • Tel: +44(0)1305 260068

ALSO FROM JULIAN EDGAR

Want to modify, restore or maintain your car at home? This book is a must-read that covers the complete setting-up and use of a home workshop. From small and humble to large and lavish – this book shows you the equipment to buy and build, the best interior workshop layouts, and how to achieve great results.

ISBN: 978-1-787112-08-7
Paperback • 25x20.7cm • 160 pages • 250 pictures

For more information and price details, visit our website at www.veloce.co.uk
email: info@veloce.co.uk • Tel: +44(0)1305 260068

Want to improve handling, straight-line performance or fuel economy? In that case, you'll achieve best results by modifying your vehicle's aerodynamics. This handbook is a must-read that takes you from testing the standard car through to making sophisticated aerodynamic modifications that have real impact.

ISBN: 978-1-787112-83-4
Paperback • 25x20.7cm • 240 pages • 443 pictures

For more information and price details, visit our website at www.veloce.co.uk
email: info@veloce.co.uk • Tel: +44(0)1305 260068

INDEX

Acceleration testing 18, 19
Accelerometer testing 10
Aerodynamic flows 45, 46
Aerodynamic pressures 49
Aerodynamic testing 45-54
Air/fuel ratio meter 55
Air temperature, intake 25
Airboxes 29
AirTabs 47
Attached flow 45
Automatic transmissions 24
Averaging circuit, ride height 53

Brakes, modification 65
Brakes, testing 43
Bump steer 37

Catalytic converter 34

Damping 40
Downforce (see lift)

Engine management 64
Exhausts 31-35, 63

Flow testing, intakes 27

G units, defined 13
G-curve accelerometer 10
Gear change points 23
Graphing, deceptive 20

Handling 36, 37
Height sensors 51

Impact harshness 40
Instantaneous acceleration 11
Intake systems 27-32, 61

Knock detection 55-57

Lateral acceleration 39
Lev-O-Gage accelerometer 11
Lift, aerodynamic 47, 51

Magnehelic gauge 28, 49
Manometer 28, 29
Modifications 60-66
Mufflers, exhaust 34

Natural frequencies, suspension 40

Oversteer 37

Performance testing 18
Power curves, measuring 9-16, 58
Pressure drops, intake 29
Programmable engine management 55-59

Ride quality 40
Road testing 5-8
Roll linearity 37

Rolling acceleration times 19

Scoops 50
Separated flow 45
Silencer (see muffler)
Skidpan 39
Sparkplug reading 56
Standing start acceleration times 21
Stopping distances 43
Suspension, modification 66
Suspension, natural frequencies 40
Suspension, testing 36-43

Throttle steer 37
Torque converter 24
Torque curves, measuring 9-16
Turbo boost gauge 22
Twitchiness 37

Understeer 36
Undertray 50, 53

Variable systems 58
Vents 49
Vibration app 41
Vortex generators 47, 48

Wake 47
Wool tuft testing 46-48